UX Lifecycle

UX LIFECYCLE

The Business Guide To Implementing Great Software User Experiences

CLIVE HOWARD
&
JEREMY BAINES

MERCURY LEARNING AND INFORMATION
Boston, Massachusetts

Publisher: David Pallai
MERCURY LEARNING AND INFORMATION
121 High Street, 3rd Floor
Boston, MA 02110
info@merclearning.com
www.merclearning.com
800-232-0223

C. Howard and J. Baines. *UX Lifecycle: The Business Guide To Implementing Great Software User Experiences.*
ISBN: 978-1-50152-152-2

Library of Congress Control Number: 2023944141
232425321   This book is printed on acid-free paper in the United States of America.

Our titles are available for adoption, license, or bulk purchase by institutions, corporations, etc. For additional information, please contact the Customer Service Dept. at 800-232-0223(toll free).

CONTENTS

PREFACE

This book is for business leaders looking to embark on the journey towards building software that creates better business outcomes through delivering great experiences.

Over the last decade one of the biggest trends in technology has been a growing appreciation for *User Experience* (UX). Perhaps the quintessential example of this is Apple. With its iMac®, iPod®, and then iPhone®, the company fused elegant hardware with intuitive and engaging software to create highly coveted products. Companies such as Amazon, Meta (Facebook), and Google have also raised the bar in terms of what people expect from the experience of using a website or application.

Today people expect more than a confusing, bland, and unengaging grey screen experience that symbolized software for much of the 1980s and 1990s (and probably still does for many employees). As the ability to create software has become easier and more accessible, the differentiator between applications is no longer their features but the experience that the applications provide. For businesses this differentiation translates into better profitability and productivity.

72% of customers abandon purchases due to poor UX (Red Signal [RS00])

EXPERIENCE HAS ALWAYS BEEN A SUCCESS FACTOR BUT NOW IT IS CRITICAL

The recognition that experience can be a critical success factor in technology has grown in recent years, but it is not new; it has been a core principle for Apple since the company's inception in 1976. Microsoft realized decades ago that making Excel® more usable reduced support calls and saved money. Perhaps more surprising to some is that IBM saw the value of user experience in the 1950s when it hired renowned industrial designer Elliot Noyes. It was IBM's second President Thomas Watson, Jr. who espoused, "Good design is good business." However, as we will discover later in this book, UX is much more than design.

> "I think there is a profound and enduring beauty in simplicity; in clarity, in efficiency. True simplicity is derived from so much more than just the absence of clutter and ornamentation. It's about bringing order to complexity."

> Jonathan Ive, Chief Design Office, Apple

As the pace of work increases and the demand for real-time insights and actions grows, the software that enables business functions needs to be better. Technology entrepreneur Marc Andreessen wrote in the *Wall Street Journal* [WSJ00], "Software is eating the world." This encapsulates the drive within organizations both large and small, and whatever their industry, to become technology businesses built on and delivering software solutions. In that world, the company with the best software wins and the best software is that which provides the best experience.

In a study, "Exploiting the Software Advantage" [Freeform00], looking at digital transformation within the enterprise Freeform Dynamics and CA Technologies found that the top three digital aspirations were improved customer satisfaction, better workforce productivity, and improved customer experience. All of these have a close relationship with UX, as we will explore.

The phenomenal growth of the internet in the early part of this century both democratized software development and brought UX to the forefront. For many years now, e-commerce companies have labored over every detail of their website's experience to drive growth and revenues. They have learned that the smallest change to text or color, or a slight improvement in page loading time, can have a remarkable impact on sales. For example, Amazon quickly learned that the color of their "buy" buttons could drive more purchases.

> *81% of executives said that providing a personalized customer experience is among their organization's top three priorities* [Optimizely00])

BEYOND THE CONSUMER: UX DELIVERS VALUE FOR BUSINESS APPLICATIONS

It can be natural to think that UX is only for consumer-facing software, but perhaps greater value can be found behind the company firewall.

As the attention to experience spread from e-commerce sites to social media, and mobile, people became used to having great experiences when using software. This has increasingly led to them expecting similar experiences at work. Traditional workplace productivity applications (e.g., billing and invoicing, materials management, Customer Relationship Management, and expenses) in many cases fall short of the modern website or mobile app. Simply having software that delivers on functional requirements is no longer enough.

We see this in the world of shadow IT, where users bring their own software to use at work. They would rather use Dropbox than the corporate file share. Box.com, another file sync and share company that targets the enterprise, touted UX as its competitive differentiator from the very start.

People gravitate to these modern solutions even when they provide fewer features than traditional products because they offer a highly intuitive, efficient, and satisfying experience. It is not surprising that these new software organizations are disrupting established markets and creating new ones. That workers can and do choose such experiences over the official company offering means that organizations need to improve their own.

MANY BENEFITS THAT MAKE FOR A BETTER BUSINESS

The enterprise and other software providers such as independent software vendors (ISV) are realizing the benefits of UX in their Business-to-Business (B2B) or Business-to-Employee (B2E) software. B2B is where a business builds software that is used by another business, and B2E is typically where a business creates software for its own workforce to use. This contrasts with Business-to-Consumer (B2C) applications like Facebook or iTunes that people use at home. This is not because they simply want software that appeals to employees and partners, but they see that better UX means better business.

To understand why this is so, consider that if software is easier to learn and use, and is faster and more reliable, then there can be multiple benefits:

- Training and support costs are reduced
- Task times are improved
- More tasks are completed
- Outcomes can be more accurate with fewer errors
- Business processes can be better defined and followed
- Improved workforce satisfaction

The result is a more competitive, productive, and potentially profitable business. This is visible at GE through their investment in UX Centers of Excellence that are delivering billion-dollar benefits. The company stated that "[UX] is a necessary competency to navigate today's competitive landscape." And two globally established and highly successful management consultancies, Accenture, and Deloitte, acquired digital agencies in part for their expertise in UX. The revolution in mobile apps is accelerating enterprises' belief in UX; for example, Barclays Bank adopted a user-centered approach for the first time when delivering its first mobile banking apps.

The only thing more expensive than writing software is writing bad software.

Alan Cooper, Founder, Cooper

AFTER RECOGNIZING THE VALUE OF UX THE QUESTION IS HOW TO REALIZE IT

The challenge for many organizations is that having recognized the potential of UX they do not know how to go about delivering it. What exactly is the scope of UX? How does a company improve the UX of the software solutions that it produces? What skills are needed? What impact does it have on the software development process? What is the effect on the wider business? These are all questions that are heard time and again from business leaders who want to embrace UX.

We have seen firsthand how increasingly important UX was to the overall success of any software project. This is why we focused our business on delivering great user experiences across B2C and B2B/B2E solutions. Since then, we have each continued to grow our knowledge and experience of UX through both delivery and learning from the stories of others.

The UX Lifecycle was born of real-world experience and applied to help our clients. It provides a methodology framework for implementing continuous improvement through UX within an organization whether large or small. This book will not make you a UX practitioner (that's a specific set of skills and experience) but instead will help you begin the journey that will put in place the business case, education, processes, skills, tools and, perhaps eventually, the philosophy to deliver great user experiences. These in turn will drive success in the modern software enabled organization.

REFERENCES

[Freeform00] Freeform Dynamics and CA Technologies, *Exploiting the Software Advantage: Lessons from Digital Disrupters – Rewrite*, available online at https://docs.broadcom.com/doc/exploiting-the-software-advantage 2016

[WSJ00] Andreessen, M., *Why Software Is Eating The World*, Wall Street Journal, available online at http://www.wsj.com/articles/SB100014240531 19034809045765122509915629460, 2016

[RS00] Red Signal, *72% of the Customers Abandon Shopping Cart Sales Due To Bad User Experience*, available online at http://www.redsignal.net/blog/72-of-the-customers-abandon-shopping-cart-sales-due-to-bad-user-experience/, 2012

[Optimizely00] Optimizely, *3 Technology Trends That Are Transforming The Customer Experience*, available online at https://www.optimizely.com/insights/blog/technology-trends-transforming-the-customer-experience/, 2015

Clive Howard
Jeremy Baines
September 2023

CHAPTER 1

THE AIM OF THIS BOOK

WHO IS THIS BOOK FOR AND WHAT YOU WILL LEARN

This book is for business leaders, senior management, board members, and those that set strategy for software solutions. It is also for those who want to influence senior leadership to sponsor a UX program. The organization could be an enterprise, a small or medium-sized business, Independent Software Vendor (ISV), or startup. UX professionals may find it useful if looking for a methodology, especially one that operates at scale.

Implementing UX requires change that is best sponsored from a senior position. That is because when UX is done correctly, it will touch many parts of an organization, not just those directly responsible for building an application (such as development teams). This approach often requires cross-functional cooperation that may only be possible with support from leadership.

This book will equip you with the knowledge to help put in place a UX program. To do that, we will address the following key questions:

- What value does UX deliver to a business?
- How is that value realized?
- What is required to realize that value?

In the chapter "What is UX?," we define UX by examining the disciplines involved and the important principles from which the benefit is derived. In "Making the Business Case," we explore the value drivers of efficiency, effectiveness, and satisfaction that help to shape a business case. In "Getting Started." we consider how UX and the case for it can be sold to the business and how to identify the key stakeholders. Then, in "Roles and Skills." we discuss the skills, roles, and tools needed.

We define the central role that users play throughout any UX implementation. We then model implementation through the UX Lifecycle.

There are two primary ways to improve the design of a company's products:

1. Have a CEO who makes product design a priority

2. Have an executive team that makes product design a priority

Delivering UX to New Places

UX will increasingly be relevant to parts of organizations that it has historically not been. For example, IT functions may not have embraced UX in the past, but this is going to change. This is for several reasons.

For many people, UX will be most familiar through Business-to-Consumer (B2C) projects such as company websites or customer-facing mobile apps. These are usually delivered through marketing functions, and often by using outside agencies with UX skills. These applications are growing in number and complexity, and businesses find they need to deliver more apps faster. As a result, they are increasingly taking development of such projects in-house and so also need to adopt UX capabilities.

Where UX has traditionally been ignored is with Business-to-Business (B2B) or Business-to-Employee (B2E) software. That is, applications that serve business users for the purposes of work, often referred to as productivity applications. These are typically developed by internal IT teams or ISVs (companies that build software). They are environments that have traditionally not embraced UX but that will now need to. This book should help them to do that.

Addressing Scale and Complexity Requires a Methodology

The *UX Lifecycle* provides a methodology framework for delivering UX. Using this framework, each organization will need to find the processes that work best for them, and UX practitioners will have their own unique contributions to make. It will provide a starting point from which businesses can build their own best practices.

As we move into an increasingly software-driven world where we have applications growing in both number and complexity, we need to look at how UX can be scaled (an area where the UX Lifecycle helps). For example, an organization may have one or two mobile apps for customers, but can have many tens or hundreds for use internally. Increasing complexity comes in the form of technology requirements and in the delivery paradigms from mobile to wearables to the Internet of Things.

Fifty-three percent of enterprises that scored as digital disruptors also now see themselves as software companies [Freeform00]

With that change comes added scale and complexity in software development that UX will need to address, and this book will consider ways to do that. Many development teams are already being pushed to change to deliver innovation faster and UX can be part of that.

UX cannot be merely added on to the development process by a small external team. To deliver the type of experiences that change a business requires a strong internal commitment to the right skills and processes. This book and the UX Lifecycle will prepare you to take that journey.

At the end of this book, there are two example case studies that provide a guide as to how UX adoption may look within large and medium-sized organizations.

DEEP DIVE: HOW DOES UX DIFFER ACROSS B2C, B2B, AND B2E?

While UX for B2C has much in common with UX for B2B/B2E, there are substantive differences that range from practical to cultural. These can be important to understand, especially when building a business case and assembling implementation teams.

B2C software, such as company websites or mobile apps, is usually driven by stakeholders with experience in optimizing for better outcomes through changes in the user experience. They have not always practiced UX, but the idea that there can be value in improving the user's experience is already there. The functions within the business responsible for such solutions, typically marketing, may have easier access to the type of skills required to improve UX.

B2E and B2B applications tend to address more complex business processes and are the product of technically-led teams, most often in IT. These environments typically have less experience of optimizing software experiences and can lack the know-how to realize and demonstrate the value in doing so. They can also find it challenging to find the right human resources to address UX.

Both B2C and B2E/B2B can be important for businesses and this book can help implement great experiences across all software solutions.

REFERENCES

Kleiner Perkins Caufield Byers. *Design in Tech Report. 1st ed.*, available online at: *https://designintech.report/wp-content/uploads/2018/11/designintech2016_small.pdf*, 2016

Freeform Dynamics and CA Technologies. *Exploiting the Software Advantage: Lessons from Digital Disrupters – Rewrite*, available online at *https://docs.broadcom.com/doc/exploiting-the-software-advantage*, 2016

CHAPTER 2

WHAT IS UX?

MUCH MORE THAN A PRETTY INTERFACE

Perhaps the most significant challenge when considering UX is how widely misunderstood the term is. In this chapter, we define what we mean by UX. We do this in the context of skills, steps, principles, and workflow. Looking at skills helps in understanding what UX is, as is elucidating the relationship between the key disciplines that are involved. Beyond skills, UX is a set of steps that begin with identifying the objective and the stakeholders involved, including the users. There are some key principles, such as consistency and familiarity that UX teams apply to realize the benefit that UX offers.

Fundamentally, a great experience is built on having an underlying workflow that users follow, which is optimized for the most desirable outcome. Software then enables this workflow, usually through some form of User Interface (UI), in a way that must provide a great experience. Often, UX experts talk about their work in emotional terms. Empathy is a commonly used word. It is true that UX aims to appeal to users' emotions and the process of creating better experiences seeks to understand users' feelings. However, this type of terminology can make it challenging to sell UX in a business context. We therefore tend to focus on ways of positioning UX that will appeal more to those managing the finances.

HOW DOES UX RELATE TO THE WIDER BUSINESS CONTEXT?

This book is focused on UX and how it concerns the business, different roles, and the end user. However, it is important to note that this focus should not be taken to mean that UX is (or should be) the "be all and end all." There are, of course, other, and wider concerns at all levels.

For example, various people involved in UX, such as product managers or business analysts, will have other important considerations that will impact UX or lie outside of it. The budget is a clear example of this, where the available financial resources need to be managed for various concerns. People in technical development roles, such as software architects, will need to factor in many issues that fall outside of UX when designing solutions.

UX ownership can be particularly challenging. A product owner may naturally think that they own the UX of their product. Suggesting that this moves to the UX team may appear like they are then secondary to that team. This is a delegation of responsibility, and the product owner retains overall ownership but simply passes the delivery of UX to that team, just as they would pass development to the development team.

Although we do believe that UX should be a priority for all those involved in the creation of software, it is not the only priority. Balancing priorities is clearly a challenge that each organization will need to address.

DESIGN, RESEARCH, SCIENCE, AND DEVELOPMENT: A MULTI-DISCIPLINE PRACTICE

Defining UX can be difficult because it spans multiple disciplines and a UX team will often pull in a breadth of skills from arts to science. There is a very common misconception that UX is visual design. Visual design may form part of UX, but it is much, much more. A good analogy for UX is buying a car. One does not buy a car solely based on how it looks. Most buyers will want to know about considerations such as comfort, handling, ride quality, features (e.g., GPS), fuel efficiency, and insurance costs. All these considerations together form the experience of owning a car, not just its appearance.

Software UX is similar in that it has many considerations:

- How easy is it to understand what actions to take?
- Does it respond quickly?
- How quickly can a task be completed?
- How reliable is the application (does it "crash")?
- Does it seem secure and is my data safe?
- Is the outcome accurate or are there errors?

These questions are important to address because so often companies believe that UX can be done later (after development) or that it "just" requires a graphic designer (perhaps by borrowing one from marketing). UX is not just visual design! One person may be able to deliver UX in certain circumstances, but the key skillset will not be graphic design.

This confusion arises for various reasons. One is that usually those in the creative design industry talk the most about UX, although they often do not truly practice it. Another is the use of the term "design," such as in "design thinking" or "user-centered design," which people assume to mean visual.

Another popular misconception about UX is that it is part of the software development process. In fact, the development process is a part of the UX Lifecycle. UX needs to start before any features of the software have been decided, and it continues beyond the application being deployed. UX must be baked into the entire process of creating or changing an application. Actual development, or coding, is part of the UX process, which may seem controversial to some but is critical to understanding the full breadth of UX.

As Much Science as It Is Art

Core to any UX program is research; without research, there is no UX. It is important to recognize that research is never as easy as simply asking the user what they want or how something could be improved. Often users cannot articulate this but the skill of the UX professional is to unearth it through research. Research involves numerous methods, some of which are discussed later in this book.

Not everything will be unique to an application. Some research has already been done, and some best practice and conventions already set. There are ISO standards (e.g., ISO 9241 or ISO 13407) with respect to usability. Usability is a child of UX; however, the two terms are often conflated. Usability is a vitally important element of an overall experience.

There are existing conventions. For example, the icon that makes text bold in a word processing application is well-established and no one would think to create a different one. There is also detailed research on how best to align the labels and input fields on a form or the optimum time it should take for an application to respond when an action is started.

ISO 9241 definition of usability:

Extent to which a product can be used by specified users
to achieve specified goals with effectiveness, efficiency, and
satisfaction in a specified context of use.[ISO00]

Before UX became closely associated with visual design, it was mainly
the preserve of experts in the field of Human-Computer Interaction (HCI).
These were often individuals with skills in science, such as psychology. There
are several well-developed theories that can be applied to software user
experiences like Hick's Law (decision-making), Fitts's Law (movement) or
Fogg Behavior Model (behavioral reasoning). Then there are heuristic principles, such as those outlined by renowned usability expert Jakob Neilsen.

Established and proven research and standards like this can help to
shape the fundamentals of good experience without having to go through
trial and error. By embracing existing research, a UX team can focus on
what is unique to your application and users. This makes for a better experience and saves time.

DESIGN PHILOSOPHIES

Design Thinking

Design thinking, which has its roots in the 1960s, has become a popular approach to implementing UX. GE and IBM have adopted it for this
purpose, but they (and others) also use design thinking as a way of business process engineering in scenarios where no visual design is involved.
While the use of the term "design" may suggest otherwise, the principles
of design thinking do not require visual design.

User-Centered Design (UCD)

User-Centered Design is another approach applied in UX scenarios.
Visual design teams often embrace this method, which provides a set of
processes that focus attention on the user. However, the results do not
have to be reflected in visual design since UCD, like design thinking, can
be applied to create any type of process (such as a business service).

Software Design

Software design refers to the requirements of an application, and how
it addresses those requirements programmatically and in terms of the

relationship between systems. If discussing design with a software architect, it can be useful to be clear about whether one is referring to the interface and experience or the functional requirements and architectural ways in which they are delivered. For example, a design pattern in software development references a reusable approach to solving a technical requirement in a programmatic way.

FIVE STEPS TO CREATING A GREAT EXPERIENCE

Step 1: The Objective

Identify the objective that needs to be addressed within a software solution and that relates in some way to the experience. The objective may be the purpose of the whole application or just a small feature. It is often a balance between user needs and business requirements. Some make the argument that software features should always be driven by the user. In some cases, this will be true, and the user has needs that must be addressed. However, the business exists in its own competitive environment outside of the software and will have its own requirements.

Sometimes, the objective may not be obvious and generative research can be done to identify it. *Generative research* is the study of users or an existing system that identifies problems or areas for improvement. Research is fundamental to UX and can be applied at all stages. Recognizing the symptoms of poor UX is one of the first challenges in creating a better UX. In the "Getting Started" chapter, there is a section on "How Do I Know if I Have a UX Problem?" One often needs to study users and behavior to understand where problems or potential future problems lie.

Step 2: The Process

Look at the underlying process that a user needs to complete or take part in to meet the objective. A process might be finding the best home insurance deal on a website or generating a purchase order within a business. It is important to give some focus to the business outcomes and what parts of the business will be affected.

Considering business outcomes may sound more like the solution to a B2B problem and perhaps overly grand when the changes in UX may be small. Inevitably, though, almost all UX is driven by a business outcome. Amazon is B2C, but increasing sales would be a business outcome. For taxi app Uber, which is also B2C, increasing the number of journeys a user takes is a business outcome. Even small changes in UX, such as altering the color of a button, are fundamentally driven by trying to realize a positive result for the business and creating a positive experience for the user.

[There was a] $12 million increase in profits at travel firm Expedia just from removing an optional field from a form [Econsultancy00]

Step 3: The Stakeholders

Whether the outcome is increasing customer sales or improving employee efficiency, multiple parts of the organization may be involved. Part of this requires bringing together the key stakeholders in the process, not just the stakeholders in the final application. Often the application development process will only include those most directly involved, such as IT and the product manager. This potentially misses out other important parties.

In an enterprise scenario, this may involve one or more business functions such as sales, marketing, or customer support, or even external parties like partners or suppliers. Collaboration between application stakeholders is critical to success.

Step 4: The Delivery

Attention then turns to how best to deliver the process to the constituency that will be following that process: the users. Research data is collected about users to build a picture of them and this data forms the "persona" (a term we will explore later). The aim of this is to understand the user by identifying their needs, limitations, thoughts, behavior, and context. Context has become increasingly important, as users may no longer be just sitting at their desk in an office but could be out in the world using a mobile device.

This early research supports the functional (and non-functional) requirements of the application and how those requirements are delivered in the form of an application.

Involving users at this stage may seem strange for people used to more traditional software development processes. It may make more sense to collect functional requirements first and then consider the users.

However, this often leads to requirements being missed. These may not be obvious to those involved in the requirements-gathering process who quite often will never actually use the application.

This research will inform the design of the process, experience, and UI. This is an iterative process whereby evolutions of design can be tested, and further research done.

Involving users early and often will help to avoid omissions and mistakes. Frequent research allows teams to drill into very detailed aspects of the experience.

Step 5: The Result

Once the application has been deployed, further research helps to understand how effective that delivery has been and where it can be improved further. We should also be able to prove the level of success and so the return on investment (ROI). It is this approach of "discover-build-measure" that gives us the UX Lifecycle.

Think Outside of the Application

UX can extend beyond the application itself to include user support and other user communication channels. It is popular to add the ability for users to provide feedback via the application. Online services such as UserVoice or Hotjar allow users to post feedback on which other users can vote and discuss. This can provide additional research data from which the UX team can draw.

The ability to provide feedback and the responsiveness of support services affect a user's view of an application and so have an effect on the experience (this is somewhat like how your car dealer contributes to the overall experience of owning the car).

Main Principles of UX

UX is built on research and the outcome is the creation of efficient and effective processes enabled through software to meet an objective and improve business value. We will discuss efficiency and effectiveness in the chapter "Making the Business Case." Let us now consider how UX achieves this by understanding some core principles.

Consistency: Does an application work in a consistent way?

If something is always the same, then people become used to it very quickly. For example, if the page numbers in a book are consistently in the bottom right on every page, then finding a page is quick. If the page number moved around from page to page erratically, then finding a page becomes much harder and more time consuming. Consistency enables users to work more quickly and accurately.

Familiarity: Are there ways in which an application works with which I am already familiar?

If someone picks up a remote control for a TV anywhere in the world, they can turn it on with little difficulty. That is because the universally recognized red button and power icon are the same on every continent and in every language. The same is true in software. The best mobile apps are usually those that follow the same conventions as the phone's operating system (OS). This is because the user is familiar with how the OS works and therefore, they instinctively bring that know-how into the app. Familiarity enables users to work more quickly and accurately.

Expectation: Does this application behave in the way that I expect?

When you picked up this book, you expected to start at the front and read through to the back. Had we started this book at the back or middle, then you would not have expected that and been confused. It would have taken you some time to understand that the material was ordered in a way different from your expectations, and that would have slowed you down. In software, we come to expect certain behavior or visual cues and, if our expectations are met, then we work faster. Expectation enables users to work more quickly and accurately.

Trust: Can I trust an application to be accurate and secure?

When using accounting software, you want to feel that the calculations it does are accurate and that the information it stores is secure. If you doubt the results or have concerns that your data is not being stored or it is not safe, then you will probably not use the application. You will start checking the results manually, which adds time to the process that the application was meant to make quicker. Trust makes it more likely that users will want to use the application.

Simplicity: Is the application simple to learn and use?

Perhaps the most obvious design principle is the adage of "Keep it simple, stupid" (KISS), and yet it is probably the most elusive of UX principles. When a task is simple, it is easier and faster to complete, and the outcome is likely to be more accurate. Just consider how Apple reduced the numerous controls of the typical MP3 player down to just one wheel. This interaction paradigm reduced the initial learning curve because once you understood the basic principle, then any part of the system was easy to access. Simplicity enables users to work more quickly and accurately.

When considering these principles, it is important to note how they interact with any specific implementation discipline, such as visual design.

APPLYING KEY UX PRINCIPLES WITHOUT VISUAL DESIGN

One could implement a Web API (application programming interface) with no user interface that is used by software developers when writing code and still apply the principles of consistency, familiarity, expectation, and trust. The approach for building a Web API for a software developer includes the answers to following design questions:

- Is each call to the API made in a consistent way?
- Is the feedback as the developer would expect it to be?
- Are conventions, such as authentication, similar to those of other APIs the developer may have used?
- Do they feel that they can trust the results and that any data being passed is secure?
- Is it simple to learn and use?

A Web API can provide a great user experience but has no visual design involved. There are ways of implementing UX in almost any discipline, but in this book, we are interested in software.

THE BEST EXPERIENCE STARTS WITH THE BEST PROCESS

The aim of UX is to improve the efficiency and effectiveness of users carrying out a given task; in other words, UX improves user productivity. UX can also help to channel users toward a particular outcome, such as encouraging them to buy something. However, this too can be seen as a feature of productivity as its ambition is to make the most efficient route to an outcome that provides increased value.

Productivity could be viewed as a purely B2B concept, but it is universal to UX whether B2C or B2B. An ecommerce site aims to make the process of buying as effective and efficient as possible; essentially, its goal is to make the customer productive. Consider how Amazon's "1-Click" ordering makes placing an order so easy and as a result, drives more sales. Most business software exists to enable workers to carry out a business process. That process may take seconds, or it may continue for hours, days, months or even years, but you want it to be as efficient as possible.

In this sense, UX can be seen as a form of business process optimization. It begins by looking at the process and then crafting a workflow that meets the needs of the user who will be completing it. Once the workflow has been established, only then does it get applied in the context of software. Bad processes can be presented in very beautiful ways, but they are still fundamentally bad. Great UX is not about great beauty.

Too often, organizations have approached an application from the perspective of what is easiest to build from a technological point of view. This approach often includes the process or workflow a user follows that can be engineered to suit the technology rather than the user. The outcome is often not what is best for the user to complete the process and as a result, not what is best for the business. Incorrect or badly followed processes can cause delays, failure in the adoption of the software, or poor outcomes.

OTHER EXPERIENCE DISCIPLINES

Customer Experience (CX)

UX is a subset of *customer experience,* which is a far broader topic and one that touches almost every corner of an organization. CX is becoming increasingly recognized as a critical competitive differentiator. The challenges of CX are great, from breaking a company's own well-established internal silos to re-engineering business wide processes and perhaps embracing digital transformation.

Service Design

Service Design concerns the way in which organizations are organized internally to facilitate better interactions between themselves and customers or suppliers. UX is referenced in this context as well, as a component of service design.

Employee Experience

For some businesses, employee experience is as important as CX and the two can be tightly coupled.

THREE PHASES OF APPLICATION ENGAGEMENT: FIRST, REGULAR, AND EXPERIENCED

When users engage with an application, their interaction changes over time. When crafting an experience, it is important to consider the following phases:

1. First Use

This is when users first interact with an application (sometimes referred to as *on-boarding*), perhaps through a free trial or a training course. At this stage, they are simply looking to grasp the basic features, complete essential tasks efficiently and effectively, and see if it seems reliable and can be trusted.

2. Regular Use

Once a user gets past the first use phase, they may become a regular user. At this stage, they will be performing certain tasks repeatedly and so the efficiency of workflows is critical. They will also begin exploring more functionality and deriving greater value. This continues the process of learning, and the application needs to help them along this journey.

3. Experienced Use

As a user becomes increasingly familiar with an application, they require advanced functionality and higher productivity options such as shortcuts. Again, there is a process of education but also perhaps the potential for customization.

The challenge is to provide the best experience for users in any given phase without negatively affecting those in another phase. Most of us are familiar with complaints about how overly complex Microsoft Word is because of the number of features. However, many of these were born out of the needs of different user groups and built up over time as users became more experienced. Trying to balance the needs of these different phases is a challenge UX constantly seeks to resolve.

There can also be special cases whereby the three phases never really exist for some users. An example is an application that is only required infrequently. Imagine an accounting application used by small businesses where the users only log in once a month or less. They may use it once and then not again for some time. By then, they will have likely forgotten all that they learned about using it and so never become experienced users.

HCI THEORIES

Hick's Law

The work of psychologists William Edmund Hick and Ray Hyman shows that the more choices a person has, the longer it takes them to make a choice. In an ecommerce scenario, this could mean that providing customers with more choice results in them *not* making a choice. One example is that when consumers were offered a choice of six jams, 30% made a purchase compared to only 3% when faced with twenty-four jams.

Fitts's Law

Psychologist Paul Fitts identified that the time required for a user to move to a target (such as a button) is related to the size of that target and the distance to it. When using a mouse, this means that placing items closer together and making them larger helps users work more quickly. The law also applies to mobile, where an example is the position of elements on the screen relative to a user's thumb movement when holding the device. In the context of an application, this can lead to improved efficiency.

Fogg Behavior Model

B.J. Fogg, while a doctoral student at Stanford University, looked at how different factors come together to determine whether a user takes a particular action. He found that for a website, the key factors are motivation, ability, and triggers. His model can help experts to identify why users may not take certain actions, and, more importantly, to alter their behavior. This is highly relevant where it is important to direct users toward a specific action.

KEY TAKEAWAYS

UX is a multi-discipline practice with skills that cover art, science, and technology. UX is also a user-centric process by which software is created and evolves. Without research, there is no UX, but there can be UX without visual design. UX is not an add-on to software development that can be added on later.

The first step in the UX process is to identify the objective(s). Objectives are a combination of user needs and business requirements. They could be anything from increasing sales on a website to enabling a new business process through software. The means by which the objective is achieved can be large or small (for example, simply improving a "Buy" button).

The next steps to creating a great experience are as follows: identify the process/workflow by which the objective is achieved; bring together the stakeholders involved in the process, including the users; deliver the best version of the process and interface; and measure the result to prove success.

Important UX principles include consistency, familiarity, expectation, and trust. These transcend any specific discipline, such as visual design, and can be applied to the underlying process that users follow. Optimizing that process is the foundation of creating a great experience.

REFERENCES

[ISO00] ISO. *ISO 9241-210:2010 - Ergonomics of human-system interaction — Part 210: Human-centered design for interactive systems,* available online at *http://www.iso.org/iso/catalogue_detail.htm?csnumber=52075,* 2016

[Econsultancy00] Econsultancy. *Conversion rate optimization: eight case studies that show the benefit of UX testing,* available online at *https://econsultancy.com/blog/63984-conversion-rate-optimisation-eight-case-studies-that-show-the-benefit-of-ux-testing/,* 2013

CHAPTER 3

MAKING THE BUSINESS CASE

THE BUSINESS BENEFITS OF GREAT UX

Implementing UX is not as simple as hiring a UX expert and asking him to make existing application screens "look better." This is one of the common misconceptions and mistakes made by software teams. There is the need for investment in acquiring the right skills and implementing new processes. However, it is important to recognize the numerous benefits, such as increasing revenues and reducing costs.

In this chapter, we focus on these benefits so that you can identify how UX will bring value to your business and sell it to those who manage the finances. This is primarily done through the key value drivers of efficiency, effectiveness, and satisfaction. While B2C and B2B/B2E can generate value in different ways, there are some common benefits. We will explore these areas to help you identify the best business case for your business.

> 200%: [This] was how much companies, recognized as using design effectively, outperformed the FTSE between 1994 and 2003. [DesignCouncil00]

As we saw in the previous chapter, UX is about identifying and addressing a business objective and then solving it in the best way possible through software. Doing this requires optimizing both the process that users (customers, employees, and partners) follow and the method of interaction (the software). The result should deliver multiple benefits that can include reduced costs

- better outcomes
- improved quality
- reduced errors
- improved productivity
- increased employee satisfaction
- better customer service
- increased software delivery cadence
- better adoption rates
- increased revenues
- reduced documentation
- shorter development times
- build the right applications

Whatever the initial objective, UX should have a measurable outcome by which to determine the *Return On Investment* (ROI). There are established benchmarks for some scenarios, such as average task completion rate, average system usability scale score, average task difficulty using Single Ease Question, average number of errors per task, and consumer software average net promoter score. However, ROI should be based on your own desired business outcomes.

This chapter focuses on making the business case for UX. However, this may form part of a larger business case for the overall application or even a broader project. In some cases, it will stand alone, as the project is purely to improve the UX of a given application.

COMMON EXPERIENCE METRICS

- **Average Task Completion Rate**

This is the average time to complete a given task. Shorter times can mean improved efficiency or productivity.

- **Average System Usability Scale Score**

The System Usability Scale is a way of measuring usability via a ten-question survey that provides a score.

- **Average Task Difficulty using Single Ease Question**

The Single Ease Question (SEQ) measures how difficult it is to complete a given task using a seven-point rating scale ranging from very difficult (1) to very easy (7).

- **Average Number of Errors per Task**
An average count of how many errors (mistakes or technical failures) occur as users attempt to complete a given task.

- **Consumer Software Average Net Promoter Score**
A way of measuring word of mouth using an eleven-point "likelihood to recommend" question (0 to 10). This is typically for B2C scenarios.

KEY UX VALUE DRIVERS: EFFICIENCY, EFFECTIVENESS, AND SATISFACTION

When looking at the results of UX, it is important to understand the fundamental benefits from which value is derived. The concepts of efficiency and effectiveness are important in UX and are referenced, usually together, time and again. You may have already noticed it in this book. So, what do we mean by them?

Efficiency

Efficiency is the optimum use of resource; often that is time: a business process, workflow, or task should not take any longer than is necessary. This is regardless of whether the person doing it is a novice or expert. Efficiency saves time and that time can translate into numerous benefits. That might be encouraging customers to buy more often (Amazon's 1-Click ordering) or an employee can complete more tasks in a day.

Effectiveness

Effectiveness is about achieving the best possible outcome from a given process, workflow, or task. Simply completing a process quickly is not beneficial if the result is wrong or has limited benefit. Minimizing errors should always be an ambition when creating software. Accuracy is an often-forgotten aspect of software development. (How many of us have filled in timesheet applications incorrectly simply because it was faster and easier than entering the correct information?) The result is that anything the business does with that data, such as costing future projects based on historical time usage, is based on misleading information, which is likely to result in poor decisions being made. Through UX, we aim to improve the workflow in a way that increases accuracy and so leads to a better output.

Satisfaction

In addition to efficiency and effectiveness, there is a third benefit that is often cited: satisfaction. If an application provides a more satisfying experience, it can become what is sometimes called *sticky*. That means people like to come back to it. They are happy to use it rather than simply compelled to because they must, such as the way in which employees are required to use a certain application.

Satisfaction can be a challenging benefit to sell as part of a business case because the results can seem less tangible than efficiency and effectiveness. However, if one considers the importance of retaining customers or generating user acceptance then it does have value. We have seen situations where employee churn was a direct result of poor software UX and reducing that churn lowered related costs (employment and training) and improved the business outcomes. Satisfaction can also be linked to user or customer engagement with the business which is a KPI for many organizations. Fortunately, if an application delivers efficiency and effectiveness, it usually also creates satisfaction.

[A] $125 million software project was abandoned by Avon Products after tests showed it was so difficult to use that salespeople left the company as a result. [WSJ00]

BUSINESS-TO-CONSUMER BENEFITS ARE OFTEN OBVIOUS

B2C UX is often discussed in the context of ecommerce where many organizations have seen tangible benefits. Figures that show a substantial increase in sales and customer retention or a reduction in bounce rates and abandoned shopping carts can easily support a business case. Such returns can make justifying UX investment easy.

When the business's bottom line depends on the number of customers taking a given action, whether it is buying, registering, signing up or something else, anything that improves those metrics should make for a solid business case. Crucially though, businesses need to have a solid understanding of where improvements can come from in the first place.

For example, it is not unusual to find businesses that sell online with no understanding of how many shopping carts are abandoned on their site. Each abandoned cart represents lost revenue. To demonstrate the value in better UX, one first needs to identify and measure the areas where improvement will have value. Identifying where poor UX is a problem and where improving UX is the solution can be tricky. (This is something we explore further in a later chapter.)

> [There was a] 50% reduction in the number of abandoned shopping carts at online retailer ASOS by dropping the mandatory account creation requirement. [Econsultancy00]

THE CASE FOR BUSINESS-TO-BUSINESS: CHALLENGING, BUT COMPELLING

The business case for non-consumer-facing applications can seem less tangible, but modern, smart businesses are recognizing the potential and realizing the benefits. GE, which has invested heavily in UX, hypothesized that it could realize $10 billion in savings across its infrastructure business alone [Forrester00]. New companies like Box.com and Xero, which have B2B products, are thriving by providing a better experience than traditional competitors.

The Impact of Bad UX

Often, badly implemented business processes or UI result in poor or incorrect outcomes because the employee struggles with the software. If the experience is complex or confusing, then users find workarounds that can operate outside of the official process. This can lead to problems, such as data not being entered or entered incorrectly, or even users bypassing the software altogether. Software is supposed to enable the business process, not lead to users creating their own ad hoc or off-grid process because using the software is such a problem.

> Sixty-seven percent of users say that bad UX negatively impacts brand image. [Forrester00]

Bad UX can manifest in several ways, the following are only a few.

Data

Sometimes referred to as the "new oil," data is increasing in volume and value. Poor UX often leads to incorrect data being entered or taking longer to input. The results can include added expense in data cleansing and decisions based on inaccurate or old data. The issue of speed is also becoming more critical as businesses attempt to make decisions in real time.

Intellectual Property

A concern for many businesses is *shadow IT*, where users decide not to use the official application and find a third-party solution. Shadow IT can occur because of the poor UX of the applications that employees are supposed to use. This creates problems, such as security or sovereignty of data. Company data may reside outside of the business in a geographically inappropriate or less secure environment.

User Training

In scenarios where users struggle to use an application, we often see high user training costs, higher turnover of staff, and increased support costs. "Ad hoc helpdesks" may be created inside an organization simply to deal with the issues that result from poor UX, which only increases bureaucracy and cost.

Software Adoption

A 2015 McKinsey study found that 70% of enterprise IT projects failed due to lack of adoption resulting from employee resistance [McKinsey00]. Most frequently, this is a result of bad UX where users simply find the software difficult to use and prefer to avoid it. The costs of such failure can be high.

Productivity

For a long time, many organizations have valued software functionality over usability. Users complaining about how long tasks took were ignored because the tasks got done. Users, especially employees, had no choice but to use the software. However, this line of thinking ignores that what users are really complaining about results in poor productivity.

Benefits of Good UX

Improving UX of B2B and B2E applications can address multiple challenges facing modern business (for example, see those listed in "The Impact of Bad UX").

Data

Through better UX, applications help users to input data with improved accuracy and reduce the costs of cleaning up bad data later. By helping users to work faster, data is then available to the business faster. These benefits enable decisions based on data to be made faster and with greater accuracy.

Intellectual Property

Encouraging users to use the applications approved by the business helps to protect valuable data. This means that IP is stored according to the appropriate governance and legislative rules.

User Training

Reducing the time required to train users means that they create value for the business sooner. This is also frequently reflected in a reduction in support costs and reduces the resources needed for training and support functions.

Software Adoption

Providing a better experience is shown to support higher rates of software adoption. This avoids wasted time, effort, and cost in software solutions that are abandoned.

Productivity

UX can drive productivity through improvements in efficiency and effectiveness of the workforce and others, such as partners and suppliers. For example, by improving the UX of its process order management system, farming cooperative Fonterra improved employee productivity by 88%, saving users 10.6 hours per day [Nicoloudis00].

> Sixty-five percent of software projects were adopted in the first year where UX was a priority compared to the financial sector average of 10%. [FrogDesign00]

For some, there will be specific drivers to improve an application's UX. In an ISV for example, the primary driver may be increased sales or to remain competitive against the market. Customer expectations of software are increasing because they are driven by the experiences that they now get daily from social media and mobile apps. This is reflected in their buying decisions, and if one application is less usable compared to another, then that can sway in which one they invest.

SHADOW IT

There was a time when the only applications that people had access to at work were those provided by their organization's IT department. Today, people can access applications on the Internet for work purposes either for free or using a company credit card. Shadow IT has exploded in the age of the cloud and can cause numerous challenges for businesses. While the tools on offer can be useful, they can also result in issues such as valuable or private business and customer data being stored in less secure or unauthorized locations.

Quite often, the reason employees use unauthorized applications is because they provide a better experience that allows them to be more effective and efficient. In many cases, the business provides an equivalent application, but it provides a worse experience. A typical example is the use of Dropbox versus traditional file sharing solutions. Organizations want to reduce the use of shadow IT, but attempts to block employees using unofficial applications frequently fail. The better solution is to embrace better experiences.

GOOD UX CREATES A MORE EFFICIENT SOFTWARE DEVELOPMENT PROCESS

There can be significant benefits within the development function that are common across B2C and B2B/B2E. The UX process typically reduces the amount of change required to an application after deployment in response to user issues. One study recorded that 80% of unforeseen fixes to software are a result of users struggling with the application. [Bossert00] Up to 60% of an application's code is dedicated to the user interface and so is 40% of the development effort. According to Smith and Reinertsen, the first 10% of a product's design process can determine 90% of the cost and performance. [Smith00] These figures show that there can be considerable savings by conducting UX research upstream rather than fixing problems later.

[There was a] 60-90% reduction in the cost of development by fixing UX issues in the design phase of the American Airlines website. [HFI00]

Removing the task of creating the application's user experience from development teams means that they can focus on areas best suited to their skills and it makes them more efficient. This type of change that enables organizations to take software from idea to delivery in very short turnarounds is integral to *digital transformation*. In fact, many of the concepts and processes covered in this book could be found as part of a digital transformation.

[There is a] $100 return for every $1 spent on UX. [Pressman00]

UNEXPECTED BENEFITS MAY FOLLOW

While a business case may center on specific outcomes and metrics, UX inherently delivers multiple benefits, some of which may not be expected from the outset. An example is a project we did for an insurance provider to improve the software used by its call center staff. The initial objective was to reduce the time it took for customer service agents to carry out routine tasks related to customer calls. The obvious benefit was in reducing call times for which they could prove a clear ROI.

The call center could have used these savings to reduce the number of staff. However, they chose to reinvest the time saved into providing a better customer experience, which boosted brand loyalty. The new easier-to-use application improved the experience for the call center staff and this reduced employee churn. That reduction, combined with a more intuitive interface, reduced training and support costs. While the initial business case was built around a single metric that achieved business buy-in for the project (12 times the ROI on that metric), the actual benefits to the business went further.

No matter what the business case outlines as the benefit, the implementation needs to include the necessary mechanics to measure it. This is often forgotten and leads to the business not being able to measure the actual ROI.

Further resources that explore ROI of UX in detail and how it can be measured include *Return on Investment (ROI) for Usability* by

Nielsen Norman Group and *Measuring the User Experience: Collecting, Analyzing, and Presenting Usability Metrics* by Tom Tullis and Bill Albert.

KEY TAKEAWAYS

UX can always have a measurable outcome by which ROI can be determined. This should be identified at the outset and the necessary capabilities put in place to measure improvement. Bad software experiences can also have a measurable outcome.

There are three primary value drivers from which the benefits of UX are derived. Efficiency is the best use of resources such as time; effectiveness drives better and more accurate outcomes; and satisfaction builds user retention and aids software adoption.

There are many ways in which UX can benefit businesses across B2C, B2B, and B2E solutions. These include reduced costs, improved quality, increased productivity, higher revenues, and reduced errors. Often, there are unexpected consequences, such as improved morale among users.

Most software projects fail due to lack of adoption and most software change requests result from poor usability. UX can address these issues and deliver other significant benefits within the software development process. Time and cost can be reduced, and developer resources can be better employed.

REFERENCES

[DesignCouncil00] Design Council. *Design in Britain 2004-2005, 1st ed.*, available online at *https://www.designcouncil.org.uk/our-work /skills-learning/resources/design-britain/*, 2004

[WSJ00] Fitzgerald, D. WSJ. *Avon to Halt Rollout of New Order Management System*, available online at *http://www.wsj.com/news/ articles/SB10001424052702303932504579251941619018078*, 2011

[Econsultancy00] Econsultancy. *Conversion rate optimisation: eight case studies that show the benefit of UX testing*, available online at *https:// econsultancy.com/blog/63984-conversion-rate-optimisation-eight- case-studies-that-show-the-benefit-of-ux-testing/*, 2013

[Forrester00] Forrester.com. *The Seven Qualities Of Wildly Desirable Software*, available online at *https://www.forrester.com/*

*report/The+Seven+Qualities+Of+Wildly+Desirable+Software/-/
E-RES58115,* 2011

[McKinsey00] McKinsey & Company. *Changing change management,*
available online at *https://www.mckinsey.com/featured-insights/lead-
ership/changing-change-management,* 2015

[Nicoloudis00] Nicholas Nicoloudis. *User experience improving employee
productivity by 88%, a process improvement case study with Fonterra,*
available online at *https://www.linkedin.com/pulse/user-experience-
improving-employee-productivity-88-case-nicoloudis/,* 2017

[FrogDesign00] Frog Design. *IPC Trader Tools,* available online at *https://
www.frog.co/press-release/ipc-partners-with-frog-to-create-multi-
media-strategy-to-further-redefine-trader-communications,* 2016

[Bossert00] Bossert, J. ASQC Quality Press. *Quality Function
Deployment,* 1991

[Smith00] Smith, P. and Reinertsen, D. John Wiley & Sons. *Developing
Products in Half the Time,* 1998

[HFI00] Human Factors International. *Usability: A Business Case,* available online at *https://humanfactors.com/downloads/whitepapers/busi-
ness-case.pdf,* 2015

[Pressman00] Pressman, R. McGraw-Hill, *Software Engineering,* 2014

CHAPTER 4

GETTING STARTED

BUILDING A SOLID FOUNDATION FOR A UX TRANSFORMATION

Establishing a business case for improving UX is only the first step. Very few organizations can go from budget sign-off straight to implementation. There are some critical steps, actions, and considerations that must occur first. This chapter will outline the most important of those steps which include educating the business as to what benefit UX brings and what it means for different people; who are the key stakeholders and the importance of collaboration between them; identifying the objective that UX seeks to address; the role of tools in implementation of UX processes; and the importance of analytics in reporting success.

Where there is an existing application, or an existing application development capability, there will be some process by which a user experience is currently created (even if it is not a good one). This should be investigated as there may be current practices or skills that can be re-used.

Perhaps it is solely within the purview of the development team or maybe it is outsourced to marketing or a third party. Implementing UX is not just putting in new processes, but potentially removing or adapting existing ones and redistributing responsibilities. There may be cultural sensitivities that need to be carefully addressed. Taking away the responsibility for an application from an experienced employee risks a negative reaction. If that employee is still needed in the new UX process, then the company should retain that employee.

It is important to remember that UX is not just about the application (i.e., the code and interface), but the underlying business process which may have its own stakeholders. The first step is often reaching out to these groups and preparing them for the change ahead.

SELLING UX TO THE BUSINESS THROUGH EDUCATION AND EVANGELISM

A UX program is a significant undertaking that can affect several people across an organization. Getting buy-in for such a project and then preparing people, especially the stakeholders, for the process is critical. There will be natural concerns regarding costs, the potential effect on project delivery timescales, and cultural "friction." Fear and resistance are typical where change is involved.

Education is vital to the success of any UX program. The first people to educate are those who manage the finances. They will need to be convinced of the need for UX, the business case behind it, and what delivers the results. At the next level, one needs to educate peers and colleagues so that they understand why UX will help them. If the initial plan is to improve a B2E application used by a particular *line of business* (LoB), then the head of the LoB needs to understand what benefit they are going to get. Education can often be done via presentations, sometimes in informal settings. One of our clients used lunch hour in the office restaurant to give such presentations.

> Eighty-three percent was the average improvement in [the] Key Performance Indicators (KPIs) generated by good UX. [NielsenNorman00]

Enthusiasm with Realistic Expectations

While education is important, there is also a degree of evangelism required. When it comes to change and transformation, people need to be enthusiastic and motivated. If they are excited about improvements to the productivity of their team or business function, selling more products, or gaining more customers, they will be more supportive.

A well-structured education program is vital in large organizations where UX will roll out gradually to different parts of the business. This will prepare employees ahead of time and address concerns early on. Using the experience of others who have gone before is also useful, and encouraging people to "evangelize" about the benefits that they have seen can greatly reduce "friction" with future participants.

UX can deliver remarkable benefits, and there are some examples of those in this book. We have regularly helped organizations to deliver significant ROI. However, it is always sensible to not over-hype UX when trying to convince others of the value. That can either make some people overly cautious ("It is too good to be true!") or risk people expecting too much too quickly and being disappointed.

Change is Good, But Challenging

If changing an application's experience makes users lives easier, they will support the change, although this may not be instantaneous. Even beneficial change can be challenging. We frequently attend workshops to listen to users discuss how poor an application is. However, when asked for any final thoughts at these workshops, there is one user who will sheepishly ask, "You're not going to change it too much, are you?"

One should also expect and prepare stakeholders for negative feedback. This will be because of change, which even when it is beneficial, can be disruptive. When Microsoft implemented the "Ribbon" in its Office tools, the leadership accepted that there would be short-term criticism for what, over the longer term, was a major improvement others have since replicated.

Once change is accepted and results realized, users can become the best "evangelists."

Users that are engaged in the UX process have the benefit of sharing their views and seeing how the result improves the application and their work. It is not unusual during a user workshop in which a new version of an application is being presented to solicit cheers as they see historical pain points being resolved. These users go back to their work environments and often evangelize the process to their colleagues. Finding the initial group of users may be challenging, but after that, you may find many more enthusiastic employees. This is useful, as new users can mean additional feedback. It is important to create visibility for any success, either through presentations or other channels, such as the internal organization newsletter.

WHERE DO I START?

Armed with an understanding of UX and what is involved in implementation, the first question that many people have is "Where do I start?" The best answer to this is to either look for a new software project, ideally one that is small, or identify a UX problem in an existing solution.

For the latter, it is important to spot the symptoms of bad UX and ideally a scenario where addressing it will realize obvious value to the business. This may not require addressing an entire application, but perhaps just part of it.

Starting with only a small part of the application allows UX to prove itself with minimum investment and change. Integrating UX into a larger change program affecting software development can be very helpful. An example of this is where businesses are migrating desktop applications to run in Web browsers. This inherently requires a significant rework of the software and provides an opportunity to improve the experience.

COLLABORATION STARTS WITH STAKEHOLDERS

Identifying the key stakeholders and bringing them together is a critical step. The UX Lifecycle is an iterative process of continuous improvement, and so stakeholders will need to meet frequently.

It is important to remember that most applications are facilitating a process (a shopping cart is a process, just as completing a purchase order is). Therefore, stakeholders are not just people who deal specifically with the software application, but the business function that the process serves. It is crucial to have people involved who understand the process, the purpose of the process, and value the outcome.

In the beginning, it is the process that the UX team is interested in and not software. If the process is not right for the business and the user, then software will not help, even if its user interface (UI) provides a great experience. Too often, the most influential decision makers in the creation of software solutions are removed from the business and the users who are most directly affected.

Typical stakeholders are listed in the following sections.

Business Stakeholders

Line of Business

Most applications will fall within the remit of a single line of business even where they may affect several. For example, the company website is usually controlled by marketing. An application used within a customer call center may be controlled by customer services. Someone from the line of business should be involved in the working group. In the case of marketing, this might be the Chief Marketing Officer (CMO) or someone designated by the CMO.

Product Manager

Depending on how an organization is structured, the product manager may also be the line of business representative. The product manager has ultimate responsibility for the application. They usually define the functional requirements coming from the business and communicate them to the development teams.

Development and UX Stakeholders

Development Team

Usually, the development team is represented by an employee with technical management responsibility for the application, such as a software architect. They need to have a knowledgeable appreciation for the technology involved and the implications with regards to time, cost, and potential challenges of any proposed work because of the UX process.

UX Team

The UX team is a stakeholder with the overall responsibility for UX. That does not necessarily mean that the team or any member of it must own the UX of a product. This can be the responsibility of others, such as the product owner or product manager. There is a difference between UX and the UX process, which is like the difference between development and the development process. A product owner may not own the development process even if they own its development. IT or the development team will decide on the process, methodology, tools, and technology but the product owner remains responsible for the product's development.

While a product owner may own the UX of a product, the UX team should always own the UX process.

Users

The users are the primary stakeholders, as this is the group that will not only use the application but inevitably decide its success or failure. Success is determined through adoption rates (which are often poor for B2B applications), or sales, where customers simply choose a competitor. The main challenge with users is their diversity. Not only are there different types of users, such as the novice or expert, but users can be internal to the organization (employees) or external (customers). Getting access to them can be difficult.

Recruiting some internal staff members who are users of an application can be challenging, but engaging with an external customer base can be a far more difficult recruitment task. Fortunately, there is an expanding number of tools and methods to help. For example, there are online services which will source users based on a given profile. They can then test a prototype or application and provide feedback. This can be a very cost-effective approach, especially compared to a UX lab.

Others

Some applications may extend across multiple lines of business or there may be business functions that would have a valuable input. For example, customer service functions will frequently field complaints or suggestions regarding the company's website or other applications. This direct feedback from users can inform the UX team. Equally, changes to the application that affect users can lead to them contacting customer services. It can be helpful for that team to have early visibility into application changes.

Enterprise organizations may have roles, such as the program manager, who owns multiple technology projects and so would need to be involved. Business analysts can be important within some organizations and serve as a bridge between different functions within existing software development processes. In some cases, there can be stakeholders from outside the business, such as partner organizations or suppliers. In larger organizations, we advocate forming working groups of senior stakeholders that can collaborate and communicate regularly.

This is not an exhaustive list, and the basic aim is to involve the people in the roles and parts of the business that are most affected by the process(es) that the application enables. The actual list will change based on the project and the business. Collaboration is an important tenet of successful UX. Remember, these stakeholders should not be seen as subservient to the UX process or the UX team, but viewed as groups who bring their knowledge and concerns to the process. Some may feel initially threatened by the introduction of UX and so will require reassurance, which can be done through education.

UX TOOLS

Tools in the form of software applications can help to enable UX processes and associated tasks. There are a growing number of such tools covering all aspects of UX. Popular tools include Figma® for collaboration between different teams; Axure™ RP for wireframes; Sketch for UI design; Camtasia® for screen capture; SurveyMonkey® for online surveys; Kissmetrics for analytics; and UserVoice for user feedback. Just about every UX activity has multiple tools to choose from.

However, UX is about collaboration and so tools that can be used by other roles and stakeholders and that integrate with the tools that those people already use have additional value.

Development is an example of where existing processes will be affected by UX and where they use several tools to enable their work. Examples are management tools, including Jenkins®, Microsoft Azure® DevOps, and Jira™. Such tools help to manage processes, assets, and tasks that are foundational to software development. If one wants to integrate UX processes into developers' lives, then using their tools is the ideal way to do that. An example is how Figma integrates with Jira (for developers) and Axure (for designers).

The right tools can be beneficial in smoothing the various handover points between different disciplines involved in the software creation process. Within the UX Lifecycle, there are points at which design assets need to be handed over to developers for them to be coded. This can be challenging, as often it means developers have to recreate a visual experience based on an image of it. The chances of that design becoming corrupted during the process are high and a common problem wherever there is a designer to developer workflow. Tools such as Anima help developers to convert Figma files into code with minimal effort and maintain the integrity of the design.

BEGINNING THE UX JOURNEY

Whether the UX team is tasked with a new product or improving an old one, the initial piece of work should cover the entire application (even though it is unlikely, in many cases, that the first deliverable will be a whole application). Although as we discuss later, in the Waterfall development methodology, the first output is a considerable amount of functionality. However, even then, applications are often built in phases. Development may break it down into something smaller that can be delivered more quickly.

In the case of a new product, a reduced version with only the basic but essential features may be the first output. This type of product is often referred to as a *Minimum Viable Product* (MVP), a concept espoused by Eric Ries in his book *The Lean Startup*.

To apply the guiding principles of UX, such as consistency, the whole application must be considered. For example, what best suits a small application with minimal functionality may not be appropriate when scaled to a much larger application. Not looking at the intended whole from the beginning may mean re-inventing parts of the experience later. This negligence breaks tenets of design, such as consistency, and causes confusion for users who must adjust to a new experience. There must be a balance because the small application must be just as good as the larger one.

Creating the best experience, like specifying an application, is like building a road: you would not start without knowing where it was going.

Choosing the Right Project

In many cases, UX will begin with a small project or a small team (perhaps one UX practitioner). Barclays Bank used the development of a mobile app to embrace a UX-driven development approach. The bank chose this approach mainly because mobile apps can be short and relatively straightforward projects delivered by a small team. They present the opportunity to start a project from scratch and, as in Barclays' case, embrace a user-centric approach from the beginning.

Mobile also requires great UX to address many of the inherent limitations and challenges such as screen size, user context, and network unreliability. The processes can be put in place that later address larger projects and larger UX teams. Barclays scaled out the practices developed for the initial mobile project to cover far more of IT. This may sound like a grassroots approach to UX, but this project was sponsored by the Chief Operating Officer.

The UX Lifecycle is not dependent on a certain scale of implementation. Putting the right processes in place from an early stage is important, even if they evolve over time.

MINIMUM VIABLE PRODUCT

The concept of an MVP was developed in 2001 by Frank Robinson, CEO of SyncDev. The idea pertains to the development of new software solutions whereby the development team creates the first iteration of the application using the minimum requirements possible.

This approach enables a working and useful application to be delivered very quickly. Based on user feedback and other inputs, the application can then be iterated rapidly into a fuller and more feature-rich version that delivers increasing value. Crucially, the first iteration and every one after must deliver value to the user. Often, developers misunderstand the concept and simply produce a stripped-down version of the application that they intend but that delivers very little value in that reduced state.

The idea of iterating a live application based on user feedback while all the time making sure it is useful and delivering increasing value is clearly akin to the ideology of UX. Therefore, the two can work very well together. MVP provides a sensible framework for creating new products inside organizations of all sizes.

UX TOOLS AND TECHNOLOGY FOR A MORE EFFICIENT IMPLEMENTATION

For many organizations, especially larger ones, the implementation of any process requires tool support. For example, there are many tools used by developers to support their practices. Project management teams have tools such as Microsoft Project®. Using tools helps to document and formalize a process that anyone can then follow or access. They also provide ways of storing and managing the many file assets that UX creates.

As UX has the potential to interact with so many parts of the organization, any new tools need to either integrate with existing tools or be easy to adopt. In the interest of reducing potential resistance to the change created by implementing UX, it is important to consider using existing tools. (For example, asking a project manager to stop using Microsoft Project® is probably not a wise decision.)

The UX team will also require their own tools to support their processes and requirements. There is a growing number of such tools available to address the various aspects of implementation:

- Research
- Wireframing
- Prototyping
- Design
- Information architecture
- Testing
- Feedback
- Analytics

Where the process interacts with other business units, the UX tools should integrate with the tools already in use (for example, integrating UX tools with existing development management and planning tools).

New UX tools will often collect and manage data from research. There can be many stakeholders involved in an application and they will all need to provide input and feedback throughout the process. Having a single repository for research and input that can be accessed by all stakeholders helps to facilitate the process and provides an audit trail. This can be important if someone needs to refer back to why a particular decision was made and view the supporting research and dialogue. All research should be available to all stakeholders.

Frameworks Can Reduce Friction and Save Time

Many of today's development tools and technologies aim to make developers more productive. There are numerous technology frameworks that support UX. Using such frameworks reduces time with respect to the UX process and development, but also enables developers to work without constant UX support. For example, developers may need to build some applications where the UX team cannot be heavily involved, perhaps due to budget constraints.

By utilizing a framework, they are more likely to deliver a better application experience. Frameworks can provide a useful UI immediately, enabling developers to deliver certain aspects of an experience without requiring much time. This helps reduce tension between UX and development teams, as the developers do not need to address every detail of the UI. Frameworks can also accelerate the design process by providing designers with a starting point.

Of course, much of this relates to UI, and UX is much more than that. Having a highly effective UI on a fundamentally bad workflow does not solve the most important problem. New UX technology should be used where appropriate. If UX is about efficiency and effectiveness, then those practicing UX should aim to be just as efficient and effective. Where specific technologies or tools advance those goals, they should be embraced.

TECHNOLOGY FRAMEWORKS

Technology frameworks have been a staple of software developers for some time. Microsoft's .NET Framework, perhaps, being the most popular example. The ambition of all frameworks is to speed up development and increase reliability by reducing the amount of code that must be written. Typically, they roll frequently requested features into easily reusable code. The highly popular jQuery library solved the laborious problem of having to write, test, and manage different types of code for different Web browsers (such as Chrome™, Edge, and Safari®). Developers only needed to write one line of code instead of many.

There are now several frameworks for different development environments (such as Web and mobile) that can easily be implemented and customized to deliver UIs utilizing best practices. These inherently address numerous UX and UI considerations, such as layout and interactivity. They also help teams to employ UX principles such as consistency, familiarity, and expectation. Popular examples of Web frameworks are Bootstrap and Foundation or Hewlett Packard Enterprise's Grommet, specifically created for B2B/B2E enterprise use.

Many frameworks can be customized to address unique requirements, including branding, and some have support within their design and development tools. This makes it easy for UX/UI designers to create experiences based on the framework components that are used by the development teams.

ANALYTICS FOR RESEARCH AND PROVING BUSINESS VALUE

Gathering data about application usage is central to UX research, so using the right tools is important. There are numerous application analytics tools available. However, as many people found with Google Analytics™ for websites, simply having the tool is not enough. To obtain the most benefits, the tools need to be implemented and configured, and the data must be interpreted in the correct way. Often, this needs to be done in the context of what is trying to be achieved by UX at any given time. That means the tools may need to be tweaked to collect the most appropriate data based on changing objectives.

This is an area where different stakeholders should collaborate, as they often need to collect and have access to different types of data. The UX team will need data for their research and others may require access to data for their own business use. The UX researchers will define the data collection metrics with input from all stakeholders.

The development team will most likely need to implement the tool in the application code. Analytics can also provide the necessary metrics to support the UX business case and prove ROI. Too often, an organization will embrace UX to achieve certain improvements, but not put in place the tools to measure those improvements. A decision on whether the exercise has been successful is left to anecdotal evidence. UX should always be able to prove its worth, and using analytics to prove success metrics is the method for accomplishing that goal.

HOW DO I KNOW IF I HAVE A UX PROBLEM?

For any organization that creates software, an important question should be whether they provide a good experience. This is something that is frequently not considered, or it can be difficult to tell whether an application provides a good or bad experience.

What are some of the tell-tale signs that should make us think that an application has a UX problem? Most users will not identify the UX as being the problem. Being able to identify the symptoms of poor UX is critical to starting the process that will improve it. Some of the tell-tale signs are the following:

- users complain about an application, for example, expressing confusion
- users contact application support frequently
- on an ecommerce site, users leave without making a purchase
- users do not complete important steps within an application
- the data collected by the application is incomplete or incorrect
- tasks supported by applications take longer than desired
- users do not use the application, but find other ways to complete tasks
- users utilize workarounds that bypass parts of the software
- customers choose a competitor's software

One must search carefully to see problems relating to applications. For example, customers may be complaining about the experience they receive when contacting your business. This is due to the employee struggling to use the application that is supposed to help them to assist the customer.

For one client, we saw employees leaving private customer data (such as credit card data) lying around the office because they were trying to work around a bad application. This had obvious ramifications with respect to customer data privacy and potential legal exposure. Always be on the lookout for the symptoms of bad UX.

KEY TAKEAWAYS

A major challenge will be identifying existing UX problems. Spotting the symptoms of poor UX is the first task in understanding that there is a problem. It is often useful to start with a small project or a very focused problem whose solution would realize noticeable value.

UX creates change. While that should lead to benefits, it can be challenging for some employees. Preparing the business through education and evangelism is essential and helps secure buy-in from key stakeholders, which is critical for success. Ensure that any successes because of UX have good visibility within the business to help build support.

There are many stakeholders in the UX process, not just those directly involved in building the application, such as product owners and development teams. Collaboration between stakeholders is a critical success factor and should occur frequently. The most important stakeholder group is that of the end users.

UX leads to new processes, skills, and technology, and it requires collaboration, data collection, and analysis. This approach will require new tools for the different role groups. Integrating new tools with existing ones should reduce the impact on current teams and processes. This should enable a more efficient and scalable implementation of UX.

REFERENCES

[NielsenNorman00] Nngroup.com. *Return on Investment (ROI) for Usability | Nielsen Norman Group Report*, available online at *https://www.nngroup.com/reports/ux-metrics-roi/*, 2008

CHAPTER 5

ROLES AND SKILLS

UX ENCOMPASSES MANY ROLES

The number of unique UX job titles has grown significantly in recent years. Often, the skills required for the role do not match the title, especially as there is still a heavy focus on visual design. UX has become a buzzword in the creative design industry, getting added to agency service lists and designers resumes. Therefore, numerous UX positions, regardless of the title, require applicants to have graphic design skills.

As a result, capability descriptions are more useful than job titles, but it is easier to use role profiles to group certain skill sets. According to the latest Nielsen Norman Group study on UX careers, 75% of respondents said that they perform 16+ UX activities [NielsenNorman00]. A smaller organization may want to hire someone whose skills include UX research, architecture, and design. Positions at larger enterprises should have clearly defined roles, especially as this can help UX to scale.

While there has been a strong bias toward design skills, there was a time when UX was largely the domain of people with skills in Human-Computer Interaction (HCI) or psychology. These skills are still valuable, especially in certain scenarios. Research is the most critical skill, since it is the ability to gather user data, interpret, and utilize it. Research should be the primary skill requirement when building teams to implement UX. Practitioners can be highly qualified in academic areas: a Nielsen Norman study showed 35% have two degrees [NielsenNorman00].

UX is a process and not a job title. The process requires different skills at different stages. Anyone joining a UX team should have a keen understanding of this, and one should be wary of anyone who says that they can do everything, from strategy to front end development. Smaller organizations or larger businesses beginning their UX journey may benefit from a multi-discipline individual. However, this approach will have limitations over time.

Individuals with a breadth of UX skills can be well suited to more senior roles, such as managing a UX team. Most importantly, one should avoid the ever-growing number of designers simply adding UX to their résumés. Any UX practitioner should be research-led, but UX skills should be identified and promoted within the cross-functional team.

UX PRACTITIONERS: UNDERSTANDING BEYOND THEIR DOMAIN

It is important that the UX team develop an understanding of the business. Some degree of domain knowledge is important in understanding what can often be complex objectives and business workflows. For example, working on an application that supports a vehicle financing company requires the UX team to develop an understanding of the complex data, terminology, legal requirements, and processes involved. We worked on such a website, and a major factor in creating the user experience was ensuring that it was legally compliant. Without understanding the compliance issues, it would have been easy to create an experience that would have landed our client in trouble with regulators.

UX team members should have an appreciation for technology, most importantly, for how it pertains to the user's experience (i.e., conventions with mobile operating systems) as well as for the back-end development. This will help them better understand the challenges that development faces and improve the relationship between the teams. It will help them create and design better experiences.

Some typical UX job roles are given in the following sections.

UX Director/UX Strategist/Head of UX

The UX director (also called the UX strategist or head of UX) is most likely to be found in a large organization dealing with many applications. The fundamental skill set is the ability to oversee the entire process and have a solid understanding of everything from research to design and technology.

This is a relatively new and emerging role, but essential where the important tenets of UX (such as consistency, familiarity, and expectation) need to be applied across portfolios of applications. Remember that a single user may use multiple applications, but each one may have a different UX team. There needs to be equal and consistent experience across them all.

The UX director also has the important job of evangelizing UX across the organization. With UX processes touching so much of the business from IT to sales, marketing, or support, a senior employee must liaise with all the lines of business. The UX director must take a holistic view across the organization from the application portfolio's perspective and identify and engage the right stakeholders at the appropriate time.

They will lay the groundwork ahead of any UX implementation, making sure that other individuals are aware of the benefits of UX, but also what will be required of them and their teams. Key stakeholders' buy-in is critical to the success of UX, and the UX director is responsible for securing this. They will also oversee the development of the UX process, best practices, and any universal collateral, such as UX/UI guides or other assets shared across projects.

Among their other tasks will be aggregating best practices from the UX teams and hiring other team members.

UX Director/UX Strategist Skills

- Management and leadership experience
- Understanding of UX research and methods
- Understanding of UX design and architecture
- Understanding of software development and technology
- Ability to evangelize and educate business stakeholders
- Coordinate senior stakeholders
- Team leadership and strategic planning
- Develop UX best practices and standards
- Business knowledge
- Hiring of other UX roles
- Ability to be persuasive

UX Researcher

The main role in any UX project is the researcher. They need to have a deep understanding of the different research techniques such as user workshops, interviews, surveys, and recordings. They must know how to apply these and when best to apply them. Most importantly, they must know how to interpret the results, which can often be unclear, confusing, or conflicting.

A researcher must know how to engage users, but also work with other stakeholders at different levels within the business who will have input. They will be able to craft and employ the necessary research processes that range from initial discovery through to measurement. Creating research assets and documentation including user personas is part of their role.

UX Researcher Skills

- Experience of UX research methods, including workshops, interviews, surveys, ethnographic study, analytics, usability testing, remote studies, lab studies, and A/B testing
- Ability to create personas
- Understanding of heuristics
- Report writing
- Ability to visualize and present research data
- Understanding of quantitative and qualitative data
- Understanding of attitudinal and behavioral data
- Ability to configure analytics tools
- Collaboration, especially with domain experts
- Business requirements gathering

UX Architect/UX Designer

The role of the UX architect (or UX designer) depends on the nature of the application, but their job is to design the structural framework of the application (often referred to as the *information architecture*). In some cases, it will be two distinct roles. In the case of large applications, a UX designer may be focused on a specific workflow and so not understand the nature of the entire application; that would be a task for the UX architect. Complex applications can be made up of many different workflows, and some of these may overlap or share significant functional capabilities.

Like a business architect or software architect, the UX architect has a holistic view. They see the application in its entirety, not just a single workflow or task. They work closely with research and are able to understand the results. Part of their role is process engineering before and after that process is implemented within an application setting. They are tasked with creating the optimum workflow.

They must ensure that the key principles of UX are maintained across the whole product. For example, if two separate workflows within the application contain steps that are identical, then the UI for each should not differ. If certain visual conventions are used, then they need to be used across the whole application so that users become familiar with them and know what to expect from them.

The job of the UX designer is to plan the fundamental building blocks of the experience. They will be crafting the workflows within the application. Their job is to utilize the research and combine that with the requirements of the business to produce an actual experience that the user will interact with. In many cases, the UX designer will work up wireframes or even working prototypes that focus on the process and workflow elements. They need a reasonable understanding of technology and visual design. There will be some crossover with the work of the UI designer, and the two roles must be compatible. A UX designer does not manage certain visual requirements, such as brand guidelines.

UX Architect/UX Designer Skills

- Ability to understand and interpret UX research
- Create information architectures and content strategies
- Understand heuristics
- Create wireframes
- Low-fidelity and interactive prototyping
- Create screen flows, sitemaps, and state transition diagrams
- Understand key technology concepts
- Present solutions and concepts to stakeholders

UI Designer

The UI designer role can be separate from UX designer, although there will be overlap and the two need to work together. The UI designer applies the visual design on top of the workflows created by the UX designer. They have responsibility for the use of color, typography, iconography, and any animation-based interactions. In some environments, there may be a specialist, the Interaction Designer. The UX and UI designers are informed by both the research done for this product as well as the industry's best practices and third-party research.

The UI designer is most likely a graphic designer with an understanding of UX disciplines, such as research and established UI best practice and standards, as well as the technology environment as it relates to the application. For example, they should not design something that developers will find impossible to implement given the technological (or time and cost) constraints.

Similarly, if the application is going to run on a particular mobile device, then they must understand the UI conventions of the software that runs on that device, such as the operating system. Apple's iOS and Google's Android operating systems have fundamental UI conventions that differ, and to maintain the key tenets of UX, the UI designer must respect these in their own application designs.

UI Designer Skills

- Graphic design
- Ability to understand UX research
- Knowledge of UI conventions, best practices, and standards
- Understand software development and technologies such as HTML, CSS, and interactive design

UX/Front-End Developer

A challenge when implementing UX in a technology-led development environment such as enterprise IT or an ISV is often a lack of appreciation for the nuances of front-end (UI) development. While the UI designer understands why form labels and input fields should be aligned a certain way for better user productivity, the developer may find it faster and easier to align them differently. Making such a change in the development process can seem insignificant to the developer, but will have a major impact on the UX.

Traditional developers will have work to do that is not related to the UI, such as writing business logic code that resides on a server or database code. To address this, a front-end developer can be responsible for writing the UI code.

They must understand the importance of following the UI design and how other technology related issues may impact on the overall UX. In many respects, they are a bridge between the design and development disciplines. They will also be responsible for optimizing that code to ensure reliability and performance.

A front-end developer can also be involved in the pre-development stage when a working prototype is created. They have the potential to create the prototype in a way that code from it can be used in the final implementation. This reduces time in production. Remember that UX makes development more efficient, as well.

Front-end developers need to understand the development environment. They can also work within the development teams (either full or part time) to help with UX implementation issues. Working part time is best, as it focuses their effort on the tasks that they are best suited for. Working full time sometimes leads to them becoming just another development resource.

UX Developer Skills

- Application development skills
- Ability to appreciate UX
- Skills in optimizing code (such as SASS or LESS)
- Understanding of best practices in application performance
- Ability to create UI interactions
- Ability to convert flat designs into code
- Ability to create working prototypes
- UAT and usability testing
- Understanding of accessibility and usability within code

UX PROFESSIONALS' SALARIES AND CONTRACT RATES

The single largest cost for UX will likely be human resources. A business investing in better UX can obtain outside assistance either by engaging a company specializing in UX or by employing UX professionals. The latter means either taking on new hires or contracting in the required skills. The business case and the level of long-term commitment to UX are important factors in determining which of these an organization chooses.

To help build a business case for UX, having an idea of the costs in this area is obviously useful. We can only provide a guide, as many factors will affect the actual cost to any given business. The User Experience Professionals Association (UXPA) regularly conducts a global survey of UX professionals' salaries [UXPA00]. The results for the 2022 survey (the latest available) showed an expected discrepancy in salaries across different countries, with the US coming out as the highest. The median salaries of the roles outlined in this book are shown in the following table.

TABLE 5.1 Median salaries of some UX jobs

UX Director/Strategist/Head of UX	$150,500
UX Researcher	$106,500
UX Architect	$116,000
UX Designer	$103,000

Contract or freelance rates were not included in the survey. However, a study by UK recruitment consultants Zebra People in 2023 does provide some data. [ZebraPeople00] Rather than specific roles, Zebra organized their data by category, as shown in the following table.

TABLE 5.2 Median rates of some UX jobs [ZebraPeople00]

Director	$1,200
Head	$980
Lead	$800
Senior	$730

The UXPA, which operates chapters around the world, is a useful resource for finding UX practitioners as well as other UX-related material.

KEY TAKEAWAYS

UX is a multi-discipline process. One should be wary of the fact that UX has become a buzzword within the creative design community, and so it is important to hire for the correct skills. The ability to conduct, understand, and interpret research is critical.

The UX team should understand the business and any specific domain relating to the application. They may not be experts, but should have an understanding, especially where the domain is complex. Domain knowledge is critical, as no one understands the domain better than the users.

Due to the numerous touchpoints involved in a UX project (multiple stakeholders, teams, roles, and business functions), someone should liaise with all of them. Such oversight of UX can ensure that principles like consistency and familiarity are maintained across large applications, multiple UX teams, or multiple applications.

Job roles may include UX director/strategist (typically in larger organizations), UX researcher, UX architect/designer, UI designer, and UX/front end developer. Practitioners often have many skills that cross roles, but a team may be required to have all of them.

REFERENCES

[NielsenNorman00] Nngroup.com. *Nielsen Norman Group: UX Research, Training, and Consulting*, available online at *https://www.nngroup.com/reports/user-experience-careers/*, 2013

[UXPA00] User Experience Professionals Association. *UXPA Salary Survey*, available online at *https://uxpa.org/wp-content/uploads/sites/9/2022/05/UXPA_SalarySurvey_2022.pdf*, 2023

[ZebraPeople00] Zebra People. *Zebra People Digital Salary Survey 2022-23*, available online at *https://zebrapeople.com/digital-salary-survey/*, 2023

CHAPTER 6

UX LIFECYCLE: INTRODUCTION

AN ITERATIVE PROCESS FOR CONTINUOUS IMPROVEMENT

Fundamental to UX is that it is never finished: it is an iterative process of incremental improvement throughout the life of the application. There is always something that can be improved through ongoing research into how users interact with the application. A Nielsen Norman Group study found that on average, each iteration improved KPIs by 22% [NielsenNorman00]. There are usually new functional requirements to incorporate into each iteration, which means the basic application landscape is constantly changing, as well. Maintaining a great user experience as the application grows is as important as improving it.

Through implementing UX for numerous clients, both large and small over prolonged periods of time, we found a pattern showing that certain activities were being repeated in order. While some of the specific tasks may change depending on budget and timescales, the basic stages remained constant. Thus, we developed the UX Lifecycle, a cyclical process of repeated value-adding tasks with the aim of continually improving the experience of using an application.

The UX Lifecycle provides a methodology framework from which an organization can tailor its own specific version. Other than the general motivation behind each stage, there should be nothing that is considered required. For example, you may choose to do A/B testing in the discover stage, but it is not required.

Many UX environments will be practicing methodologies like what is outlined in this book. We are not saying that this is the only way or that it must be followed to the letter. At the very least, it is a great way to illustrate UX in action.

Stages are More Important Than Specific Activities

The stages of the UX Lifecycle resemble the typical stages that most organizations go through when creating a new product or changing an existing one. In most cases, there are requirements gathered, followed by technical and UI design, development, and then deployment. The UX Lifecycle looks very similar, but has some key differences in terms of process specifics and skills.

The most significant differences are the importance of research and the user throughout the lifecycle. In many organizations, the user is not represented at all other than perhaps through a product manager or business leader who does not actually use the product. In UX, the user is the focus of attention.

Importantly, the UX Lifecycle can be applied to any size of project, application, and organization. It applies to different budget sizes, too, as UX does not need to be expensive. Activities such as research, prototyping, and testing can be done in various ways that suit differing budgets. This is another reason why the UX Lifecycle is broken into stages, not specific activities. One can choose whether to do low-cost paper prototyping versus more expensive coded prototypes.

Within the UX Lifecycle, we talk about the processes that the UX team and others (such as the development team) follow. In addition, we are interested in the business processes that an application enables and that requires improvement. To avoid confusion, we will refer to these latter processes as *workflows*. Therefore, a process is something that the UX team might do, whereas a workflow is something the user completes. A workflow could be anything from a simple task carried out by an employee (such as locating a customer's details) or more complex (like creating a shopping cart on an ecommerce site).

The UX Lifecycle consists of three stages that are outlined here and then explored in more detail in further chapters.

HOW DOES THE PHYSICAL WORLD RELATE TO THE SOFTWARE EXPERIENCE?

In this book, we are concerned with UX as it pertains to software applications. However, software applications exist in the context of physical devices and the world. To access an application, we use a device such as a PC or mobile phone. These devices have their own capabilities, and it is worth considering them in creating a software experience.

There is a bank in Canada whose ethos is to provide the ultimate customer experience. As much of a customer's interaction is via mobile devices, this bank must deliver a fabulous user experience within its mobile apps. Banks have the added challenge of operating within a highly regulated industry. To open an account, a customer is legally required to provide certain information, such as name, address, and date of birth. This is the type of data that is difficult to enter using a small mobile device with a touch keyboard. There is a high risk that users choose not to carry out the task or make errors.

The bank's innovative solution was to use the camera available on almost all mobile phones to enable the customer to scan their driver's license. In Canada, the license has a bar code on it which provides all the data needed to open a bank account without the user having to spend time manually entering it.

Hopefully, this demonstrates how UX is about finding and delivering the most efficient, effective, and satisfying experience and that sometimes requires thinking beyond the confines of the software. The physical environment is part of the context in which the software and user lives, and we know that context is an important UX consideration.

DISCOVER - RESEARCH AND DESIGN

This stage of the UX Lifecycle is primarily concerned with research. It is similar in many ways to a requirements-gathering stage in software development. The business needs to identify the goals for either a new software application or changes to an existing one. Unlike most traditional requirements-gathering phases, this one involves the intended (or existing) application user from as early as possible, in addition to other stakeholders. Throughout the UX Lifecycle, all stakeholders should play a frequent and active role.

Research is useful for persuading reluctant stakeholders as to the need for UX or in deciding how a particular feature should be implemented. What UX teams want to avoid is a debate based on personal opinions instead of decisions that are informed by data. If a decision maker says that X would be better than Y, then they should be able to support that assertion with data.

The first step in the Discover stage is to identify the objective(s) of this iteration of the UX Lifecycle. Objectives are a combination of requirements from the business and users. Examples are

- increasing the number of people who register for a free trial of the software
- reducing the amount of time it takes an employee to complete a business process
- increasing the number of website visitors that register for a newsletter

Research focuses on the workflow, business needs, and the user. From the beginning, we build a profile of the different types of users. Remember that it is not just UI that is being tailored to the user, but the workflow itself.

The next step is to address how the workflow should be presented to a user through the software UI. A combination of user research and existing industry best practices, standards, or guidance is used. A team can use tools such as prototypes, which can be anything from low fidelity paper-based to high fidelity working code. This process of design, prototype, and test can be repeated to iterate and refine the experience at an early stage.

This stage in the UX Lifecycle ends once a final workflow and interface have been designed.

HOW DOES A NEW ITERATION OF THE UX LIFECYCLE BEGIN?

UX is an iterative process that is never 100% done. There can always be another improvement, even if the capabilities of the application do not change. Hence, the UX Lifecycle is formed of three stages that repeat and feed into one another. The last stage of one iteration informs the next Discover stage.

Application development is also increasingly becoming a cycle of continuous improvement and new features. Just think about how often you get updates to the apps on your phone. The question is not whether there will be an update, but simply how often. For some, it will be every few days and for others, every few months.

For each iteration of the application, there will be another cycle of UX improvements. The question of who begins a new cycle or what triggers the next Discover stage is intrinsic to the organization's strategy that determines how often applications are updated.

Based on data from the Measure phase, the UX team can usually always find a reason to begin a new iteration of the Lifecycle. The team can wait until the next update to the application as determined by the business and other stakeholders. It is not therefore for the UX team or process to determine when a new cycle begins, but the organization.

BUILD - OPTIMIZE AND DEVELOP

Taking the new workflow and UI as the primary input, this stage involves the technical teams. Organizations run different approaches to software development. The two most familiar are called *waterfall* and *agile*.

Waterfall is the more traditional method that usually involves a single requirements-gathering phase followed by development and then deployment. Using this approach, software can often take months or years to create the first iteration. These long timescales are the reason why many organizations move away from waterfall to agile.

Agile is the modern approach and has grown in popularity. It tends to be the de facto method for young, small organizations such as startups, but is now found in enterprises. Using Agile, an application is built in small increments, known as *sprints*, each of which is usually only a few weeks long. In this way, something new is delivered frequently, compared to the long wait to see any usable output from waterfall.

UX is ambivalent to the development methodology and can work in any environment.

It is critical at this stage that development teams, including heads of development, are onboard with the UX process because it will have an impact on their processes. However, UX should be installed in a way that creates minimum disruption to existing development process. Done correctly, UX speeds up overall software development.

If development is resistant, the risk can be high that development teams will not deliver the experience that has been created by the UX team. A simple example of this is a lack of attention to detail with respect to visual design. Visual design in UX is not just the expression of a designer's creative talent. Almost every facet will have been thought through in detail and be based on research. If development does not maintain the integrity of that visual design, then some degree of benefit will be lost.

Sixty-three percent was how much the Teehan+Lax UX fund outperformed the S&P 500; it outperformed the Nasdaq by 37%. [GE00]

IT has an Important Contribution to Make

Compliance with internal governance rules or technology standards, as well as setting important UX criteria such as performance metrics (response times, for example), are considered in this stage. Having the most efficient workflow and effective UI is useless if an application takes too long to return a result from a user's action or does not meet essential standards criteria (such as accessibility). The issue of performance is especially important when building mobile applications because network connections tend to be slower and less reliable. There are inherent performance challenges that traditional desktop applications do not have.

This is an opportunity for development to make an important contribution. Quite often during development, technical issues arise that impact the UX. An example might be that a particular database query inherently takes longer than expected. This type of issue needs to be mentioned to the UX team so that they can find solutions. The better the relationship between the UX and development teams, the better the final product. This stage in the UX Lifecycle ends once development is completed.

MEASURE - DEPLOY AND VALIDATE

Once development is complete, then the output needs to be checked by the UX team. This is to ensure that the experience that fed into the Build stage has been delivered correctly and to address any issues that have arisen during the development process. UX testing processes (sometimes referred to as *user acceptance testing* or *usability testing*) can be integrated into the code and functional testing. Pre-agreed standards and performance metrics must be tested. Only when the UX team is finished should the new product or version be deployed.

Deployment is not the end of the process for the UX team. It is important to re-validate any key performance metrics in the live environment, as they can be different to development or test environments. This is also the time when other key metrics can be measured with respect to ROI. If the objective of the exercise was to reduce the time it takes users to complete a task, then it is important to measure if that time has been reduced in the live environment. Without such measurement, then all there is at the end of the UX process is change and the hope that things are better.

Gathering analytics at this stage is of great importance. It points to where the application is being effective and where there may be work to be done. Setting up the right analytics measurements and then interpreting the results should be a competency of the UX team. This is also the time to interact with users to confirm that the goals and objectives originally discussed in the Discover phase have been met.

KEY TAKEAWAYS

The UX Lifecycle is a three-stage iterative methodology framework that can be tailored to an organization's and application's specific needs. The iterative nature of the UX Lifecycle reflects that applications naturally evolve, and their experience can always be improved.

Research is fundamental to UX and runs throughout the UX Lifecycle, but is most prevalent in the Discover stage. All decisions regarding an application's experience should be research-driven. Research is a great way to persuade people as to the validity of both the UX process and the decisions taken.

UX is fundamentally about creating the optimum workflow that guides a user to a successful outcome both for them and the organization. Once the workflow has been designed, then any visual experience (if applicable) that is delivered by the software is created.

Only when the UX team has signed-off on the output of the Build stage should an application be deployed. However, that is not the end for the UX team; it represents the start of the Measure stage, in which the outcome is judged and where ROI should be demonstrated.

REFERENCES

[NielsenNorman00] Nngroup.com. *Parallel & Iterative Design + Competitive Testing = High Usability | Nielsen Norman Group Report*, available online at *https://www.nngroup.com/articles/parallel-and-iterative-design/*, 2011

[GE00] Slideplayer.com. *UX Central. The Business Value of UX. 2. What is UX? An innovation process that makes great products. The Business Value of UX*, available online at *http://slideplayer.com/slide/4693973/*, 2015

CHAPTER 7

UX LIFECYCLE STAGE ONE: DISCOVER

RESEARCH: THE FOUNDATION OF GREAT UX DESIGN

This stage of the UX Lifecycle focuses on two tasks: research that addresses the business objective, workflow(s), and users involved; and design, informed by research, of the workflow and the visual design of the software through which the workflow is delivered.

This chapter discusses why research is so important to UX, the role of users, and how UX teams can benefit from existing research. In addition, we will delve into the types of data that are important and some of the methods used to capture it. Finally, we will consider how this informs the design of the workflow and visual design of the application.

This stage is owned by the UX team and the key inputs are the major stakeholders; the business needs; established best practice; and any existing and relevant data or business inputs like governance rules or brand guidelines. The primary output will be research assets (such as reports, documentations, logs, and audits) and a designed workflow with UI. Once the design step is complete, this stage ends.

WITHOUT RESEARCH, THERE IS NO UX

Research is the most important skill required when implementing UX and should not be limited to a single UX role. For example, UX researchers may conduct research, but UX designers and architects should also be able to interpret it. Research should always be shared among stakeholders. This is so everyone is informed and understands the basis on which decisions are made.

Why is research so important? What is the value of the output? How is it done?

Research starts with identifying and understanding the business objective(s) that any application or workflow within an application will need to solve. This understanding should be broad as well as deep. This is worth noting, because often the immediate stakeholders in a project have a particular view and set of requirements based on their business function. However, a workflow can impact other functions or job roles that are not represented by that stakeholder group.

For example, the sales department may be the obvious and primary stakeholder in an ecommerce project, but it will surely have an impact on the accounts or logistics functions, or even external parties, such as suppliers. These are also stakeholders can affect the objective and workflow.

Sometimes research needs to be performed to identify the objective; this is often called *generative research*. It involves examining existing workflows and users to identify challenges that should be addressed or new requirements. Studying users can include looking at data generated by their actions, such as application logs or analytics or watching them in person.

Quite often, UX problems are not apparent to the producers of the software, even where the symptoms of poor UX are seen. Research is needed to identify and understand the nature of the problems. For example, the symptom is that users are taking too long to complete a task. It is for the UX team to identify the reason why. They can only do that by studying the user behavior and conducting research.

In the case of a new application, there will be no existing problems, but this makes research even more important. When starting from nothing, it is easy to create problems and so one must be led more by research.

USABILITY CONVENTIONS

When creating a visual experience for users, there will often be predefined conventions that can be reused. Think about how many apps on your mobile phone use identical UI features. This is because users become familiar with certain patterns and so re-using those patterns makes it faster and easier for them to learn how to use your app. Other common conventions come from standards. Consider the universal play, stop, and pause icons used on just about every media player (physical or software) in the world. The UX principles of consistency and familiarity are well served by conventions.

However, conventions have other sources that can be discovered through research, and these help to make an app more usable. Product designer Dan Grover wrote an insightful blog post (*dangrover.com*) about Chinese mobile UI trends. While working in China, he discovered that apps for that market had certain UI conventions that he had not seen anywhere else (the "Bullet Curtain" and t-shirt icon being examples). If one were creating an app for the Chinese market, then it would be beneficial to be aware of such conventions and use them to improve the experience.

THE UX MANTRA: USERS, USERS, USERS

UX is a user-centric process where most decisions regarding the product are driven by the user's needs, thoughts, behavior, limitations, and context. Research time should be spent on understanding these. This cannot be done by just asking users, but by using a variety of tools, techniques, and processes, some of which are outlined later in this chapter. Users often recognize symptoms of poor UX, but getting from there to a solution is a reason why UX is a multi-disciplined process. A danger is that users attribute the cause of the problem incorrectly.

If you consider application stakeholders as concentric circles, the user is the closest to the application and so resides in the middle. The direct business stakeholders are next, as they understand the business needs. From there, people become less directly involved. UX professionals may never use the product in a live scenario. Therefore, how can they have a better understanding of it than people who use it every day? One should always be wary of any UX practitioner who says that they rely on their own experience rather than engaging users and doing research.

Success or failure will be determined by the user. Where many IT projects fail due to lack of adoption it is not because of functionality or that the business stakeholders failed to understand the requirements, but because users rejected the final application.

Understanding Different Types of Users

It is very rare that an application has a single type of user. Instead, it will have different types. It's likely that people with differing levels of experience, different job roles, and different motivations will use the same application. Each user then will be in a different phase of application engagement: first, regular, and experienced. These differences need to be taken into consideration.

It is of limited benefit to tailor an application to a single user. Instead, it must meet the needs of the widest possible collection of users. Research identifies the different types and makes sure that they are represented throughout the process. This includes both the most typical user types, but also those on the periphery (less frequent users for example) that are still important. This is often done using a concept called *personas*. Personas are profiles of people who represent each different type of user. How far one goes in creating a persona is for each UX team to decide, but it can simply focus on basic features such as levels of experience and personal skills, or go as far as identifying personal traits such as "ambitious" or "enthusiastic." Personas do not replace real user research, but instead provide a reminder of who your users are and what drives their behavior.

EMBRACE STANDARDS AND WHAT OTHERS HAVE ALREADY DISCOVERED

There are standards and existing best practices within the industry. For example, the best fonts to use, alignment of form labels and input fields and use of color are all areas that have been extensively studied and best practices documented. There is existing scientific and psychological research that can be used (examples include Hick's Law, Fitts's Law, and Fogg's Behavior Model). There are ISO standards, namely ISO 13407 (Human Centered Design Processes for Interactive Systems) and ISO 9241-210 (Ergonomics of Human System Interaction).

It is important for the UX team to understand the business and the domain to which an application belongs. If the purpose of the application is to serve very specific or complex knowledge areas, then the team needs to appreciate that.

Most of us are consumers and so can have an intelligent view on how to create a useful design for a shopping website. However, if the software is for use in a nuclear power plant, then the whole team involved in its creation needs some degree of expert knowledge. It is still the users (who will have the best knowledge in many cases) who lead the experience, but it is helpful to have some frame of reference by which to engage with them.

UX practitioners cannot become experts in the domain in all cases, but should have an appreciation for its unique elements and listen to the experts. There may also be existing research within that industry or domain that can be utilized. Research needs to draw on multiple sources of data.

The insights derived from research combined with business knowledge and strategy then lead the product's invention. It means whenever any decision is made by the product team (which includes UX), it is informed and not just based on a sales-driven approach or someone's personal opinion. Even UX practitioners find that what we sometimes believe to be right is proven not so by research. That is because the research draws on those who have the deepest understanding of the workflow—the users!

ISO STANDARDS IN USABILITY

The International Organization for Standardization (ISO) has developed several standards concerning usability. The most widely cited is ISO 9241, which refers to usability across software and hardware. This was originally known as *Ergonomic Requirements for Office Work with Visual Display Terminals (VDTs)*, but was renamed *The Ergonomics of Human System Interaction*.

As with other ISO standards, the full specification is an expensive purchase and uses complex language, so most people ignore it. However, for organizations where officially recognized standards are important, it can be critical in getting UX adopted. ISO 9241 is not the only standard relevant to usability, as there are more industry or use case-specific standards including for healthcare, transport, and engineering.

Another often-cited ISO standard is ISO 13407, Human-centered Design Processes for Interactive Systems, which is now part of the broader ISO 9241 (officially part 210). Such standards can be challenging for those less familiar with ISO, but there are some very helpful guides available, such as those by Userfocus (*www.userfocus.co.uk*).

For organizations that value standards and associated qualifications, there are certifications in UX. The British Computer Society offers the BCS Foundation Certificate in UX based on the ISO standard. Globally, there are others, such as the Certified Usability Analyst or Certified User Experience Analyst certifications from Human Factors International (HFI). Various UX practitioner bodies, including the User Experience Professionals Association, support the International Usability and UX Qualification Board (UXQB) Certified Professional for Usability and User Experience (CPUX).

QUANTITATIVE AND QUALITATIVE DATA: WHY YOU NEED BOTH

Research data will fall into one of these two categories.

Quantitative

Quantitative data is based on empirical measurement and can be represented through mathematical graphs or charts. An example of quantitative data is that a task takes seven minutes to complete or that six out of ten users failed to complete a task successfully. These are definitive measures and the type of data that can be used to prove ROI. For example, the objective might be to increase the number of tasks that can be completed in a given time period. Counting the tasks completed before and then after UX improvements have been made will show an exact measurement of improvement and so the ROI can be calculated.

Qualitative

Qualitative data cannot be measured in a way that provides a numerical result. Instead, it is based on opinion. An example is where a user might say that they find a particular task confusing or that text on the screen is difficult to read. Often qualitative data can be collected in a quantitative way. For example, 20% of people said that they thought the text on the screen was too small. The data being captured is qualitative because users are giving an opinion as to whether the text is too small, but the number that have that opinion is being counted to provide a quantitative result.

While either type of data has value in its own right, it is often the case that quantitative data is better understood if accompanying qualitative data is also gathered. A simple example might be that quantitative research shows 20% of tasks are not completed successfully. This data does not say anything about why they fail. Through qualitative research, one can discover the reasons for this failure.

HUMAN DATA: BEHAVIORAL AND ATTITUDINAL

An important skill in UX research is being able to interpret data, whether it is quantitative or qualitative. This is why recruiting the right skills is important to implementing UX and simply reading a book like this (or even many of the available resources online) will not achieve the level of success that someone with UX experience will.

Behavioral and attitudinal data is a useful example. Quite often, a user will express a view (attitudinal) that is not reflected in what they do (behavioral). For example, a user may say that they very rarely use a particular application feature, but when observed using the application, they use it quite often.

There can be good reasons for this discrepancy, such as subconscious reflex or perhaps they are trying to make a point during a UX interview or workshop. Whatever the reason, researching both types of data is important. Otherwise, one risks making decisions which in practice are not the right thing to do.

Microsoft based the decision to remove the Start menu from Windows 8 on behavioral data. Users were pinning the applications that they used regularly to the task bar rather than launching them via the Start menu. Attitudinal data may have highlighted how this was in fact a poor UX decision that led to a lot of negative feedback.

UX RESEARCH METHODS AND WHY YOU NEED MORE THAN ONE

There are multiple methods for conducting UX research. For best results, use a number of these, as each can experience different types of skews or bias that can distort findings. Also, use those that best fit the budget and time available. The following sections include some common methods.

User Groups / Workshops

A sample of users, representative of the overall user base, come together to discuss a workflow, application, design, or a prototype. We refer to these users as *user champions*. The same group of user champions meets throughout the UX process. Depending on the scale of the project and the size of the user base, there may be multiple groups for a single application. This approach can avoid scenarios where more vocal users can drown out the views of others within a single group.

Interviews

Instead of speaking with users in a group, they can be engaged on a one-to-one basis via in-person, phone, or online interviews. This is done for practical purposes (geography), because they represent a niche user type, or for other reasons. This can remove bias created by dominant personalities within a group, but results in direct answers to questions because there is no process of bouncing ideas around multiple participants.

Surveys

Surveys are a good way to capture both quantitative and qualitative data from a large proportion of the user base or where direct user engagement is challenging. This can also be used to establish potential ROI or prove it later. It may also help demonstrate other value in the UX process. We discovered that in a B2B or B2E scenario, sometimes just including more users in the process improves morale (another unexpected benefit). It can also help to address complex issues where reaching consensus within a small group is very difficult. It is worth noting that large numbers of users are not necessary. Research has shown that the level of insight (and therefore benefit) can significantly diminish after five users. One tends to learn very little more from user six onwards relative to the extra effort. Surveys tend to be used where large numbers provide benefits, such as proving to stakeholders that changes were seen as positive by a significant user population.

Analytics

By using data capture built into the product, the resulting data can be queried through an analytics package for insight. This provides constant, real-time quantitative data about how an application is being used across the entire user base. In addition, it can provide metrics concerning application crashes, response times, errors, and performance data. This is not as useful as engaging directly with users, but it does provide a way of viewing user behavioral patterns at scale. This is especially pertinent in scenarios where there are very large numbers of users, such as on public websites.

Observation (Ethnography)

To best collect behavioral data, observe users in the environment in which they most often use an application. This can be done by screen-capturing their desktop or device, or by being physically present when they are working (ethnography). This can also be done in a controlled environment, such as a UX Lab, in which users can be monitored interacting with an application.

A/B and Multivariate Testing

A/B testing is a frequently mentioned technique when discussing UX research. It has been used to great effect and has some high-profile proponents, especially amongst modern companies such as Facebook and Google. It involves taking two versions of an element within an application screen (or the entire screen) and applying them alternately across different users and tracking the results. Whichever version shows the best results then becomes the default version. Multivariate testing is a variation on the theme; instead of two options (A and B), there are many more options. These methods can be effective for consumer-facing applications, but risky for B2B or B2E apps where customers value continuity. Continuity is an important UX consideration, especially in certain markets such as B2B. Inside an organization, change, however good, can create disruption which can cost time and money. That is why Salesforce still only releases updates three times a year when more consumer-facing Google pushes many more.

INFORMATION ARCHITECTURE

In some situations, notably websites, design needs to extend to content and the structure of content. This is called the *information architecture (IA)*. Users will need to be able to find and access the content that they need. There are several ways this can be done: navigation, search, or surfacing personalized content based on a user's (or other users') previous interactions. The latter has been made famous by Amazon and its product suggestions that show what other products the purchasers of a given product typically also bought.

As the volume of content within websites and applications grows, how users can find what they need becomes increasingly important. A poor search feature can mean that users do not find content that exists. How content is organized is just as critical. One example is how product catalogues are structured on ecommerce websites, which have the objective of showing a customer the product they want with the minimum number of clicks. As content grows, more advanced methods of identifying what is most pertinent to the user (and their context) must be utilized.

Netflix has found ways that enable users to discover new content that is of interest to them, without them having to trawl a massive catalogue. The company gave away a $1 million prize to find a better algorithm for predicting what customers may want to watch. This should highlight how complex UX research can get in some scenarios, and why it must utilize experts in disciplines such as mathematics, science, and psychology.

DESIGNING THE WORKFLOW, INTERFACE, AND EXPERIENCE

Based on research, the next task of the UX team is to design or redesign the workflow by which the business objective will be met. That could be a shopping cart on an ecommerce site or a support ticket system in a company call center. The workflow does not have to be large, and could even be a small part of a large workflow. For example, it might be just finding a customer's details within the larger workflow of creating a support ticket or entering payment information within a shopping cart. An objective could even be to improve small elements of a workflow (for example, there has been extensive work done by many teams to find the perfect "buy" button).

Whether large or small, the design of the workflow is the most critical factor to it being successful with respect to the desired outcome. Whenever you run into difficulty using an application or website, the problem is often not the UI, but the workflow being followed. The workflow itself may be too long, confusing, or complex.

Consider mobile commerce payment providers such as Apple Pay, where the main objective is not to improve the UI of a payment experience, but to simplify the underlying workflow. By removing the need to input payment card details for each transaction, they have made paying by phone (or wearable) fast and a payment method consumers want to use.

Looking ahead to a time when more devices are connected to the Internet and allow us to interact with them, Early Warning's VP of Design and User Research Andy Goodman spoke of the concept of "Zero UI."

Zero UI is an interaction with software that has very little or no visual component. Amazon's Echo device is an example of this. Even without a UI, a workflow remains.

The UX team therefore designs the optimum workflow, and this should mean frequent interactions with users and stakeholders to continue to research and refine it. A workflow can be tested prior to UI design by using approaches, such as wireframes.

This regular engagement with users and iterations based on research is why design is encompassed by the Discover stage of the UX Lifecycle. Some may argue that when design begins, then research ends and so the Discover stage also ends. However, the design of both the workflow and the visual experience is interwoven with research, and so both reside in the Discover stage.

Once the workflow has been designed, then visual design takes place. Again, this should be research-driven, using proprietary research and that of a third party. Where applicable, other design inputs will also be considered, such as brand guidelines. The design, which can be presented in a prototype format rather than flat images, should be tested with users. This allows for the optimization of design without any impact on the development team, which is usually far more expensive. We saw in the chapter "Making the Business Case" how work done in this stage can realize savings in development.

An understanding of many of the technical considerations of the software is important to the design process. Such issues will be discussed in the following chapter. However, it is worth noting here that there is a critical relationship between development (and technology) and this stage of the UX Lifecycle.

Both the workflow design and visual design are iterative in themselves. Typically, design would be done and then tested with users, optimized further based on feedback, and then tested again. A final design signals the end of the Discover stage, and that design enters the Build stage.

INTERACTION DESIGN

Worth noting is that the design of an application is usually seen as being flat. A screen in an application appears briefly as if it were a static image. However, it contains numerous interactive elements. Experiences are multidimensional and interactive. The term *Interaction Design (IxD)* is often applied to this aspect of a software's experience.

If one looks at modern mobile operating systems, they are filled with interactive elements that, often subtly, enhance the experience. Notice, for example, how scrolling lists rarely just stop moving, but their movement slows to a stop, or how buttons on a form can have different visual states depending on whether they have been pressed. Most of us are familiar with the spinner animation that we see in applications when waiting for a task to complete.

Interaction design is an important part of the modern software experience and a skill. There is often a high degree of nuance to an interactive experience – e.g., an animation should not be too fast or too slow. Creating such experiences and then implementing them through the development process is one of the challenges of delivering great UX. Again, research will identify what works best.

KEY TAKEAWAYS

Research should be a core skill within the UX team. Even people in those roles not directly responsible for conducting research should be able to understand and utilize it. Being able to interpret users' views is the skill that differentiates the UX team from others.

There are different types of users, and they will be in different phases of application engagement: first, regular, or experienced. Research identifies them and their needs, limitations, thoughts, behavior, and context. Users are critical to a successful outcome, and maximizing their efficiency, effectiveness, and satisfaction will determine if the overall objective(s) is met.

Research data is either quantitative or qualitative. Quantitative data is measurable in a numeric way, whereas qualitative data deals with user thoughts and opinions. Both are equally important in understanding the user. Behavioral and attitudinal data provides additional insight and understanding.

There are many different research methods that vary in complexity and cost. These can include user groups, interviews, surveys, analytics, and user observation. Each has its own benefits and potential biases, and therefore, more than one method should be used.

CHAPTER 8

UX LIFECYCLE STAGE TWO: BUILD

OPTIMIZE THE EXPERIENCE, THEN BUILD IT

The final output of the Discover stage is a designed workflow with a UI ready to be developed. In this stage, the development team implements and addresses technical considerations. This stage can be owned by development with support from the UX team. In many organizations, removing ownership of an application from IT would be difficult. In this chapter, we outline the importance of development and its relationship with the UX team, especially the issues that challenge that relationship and how it can be harmonized. This can often be an awkward relationship, particularly in the early days.

The primary inputs (in addition to the output of the Discover phase) are technical considerations (such as architecture, infrastructure, and resources) and standards (such as accessibility and security). The output of this stage will be a working application or iteration of an application.

Historically, many software experiences have been created by the development team driven by the issues of most importance to them: technology and speed of delivery. This is often reflected in the overall experience. Examples are

- When asked why the steps in each workflow are not in the order that makes sense to users, development will answer that it was because that order was easier to code.
- When some UI elements are located in a counterintuitive place, the team explains by saying that was where there was some empty space on the screen.

This is not a criticism of developers, simply a reflection of how their traditional responsibilities have not included delivering the best UX. (They should not be expected to have UX skills.)

Development teams have been judged and remunerated largely on delivering functional requirements. How those requirements are delivered has not contributed towards their KPIs. Hence, many applications do what they are functionally supposed to, but in a way that is not the most efficient, effective, or satisfying from a user perspective. The result is that the application does not serve the business as well as it could or fails to be adopted, which can mean writing off the project.

TECHNOLOGY LIMITATIONS

Technology choices also bring with them potential constraints and limitations that could affect what can be done from a UI perspective. Some technologies have significant limitations (such as no ability to use color), while others bring existing experience conventions with them (such as mobile operating systems). The UX team needs to consider this in their work.

Fundamentally, technology limitations should not be seen as an impediment to UX, but just another consideration. Great UX can be delivered via a green-screen application. However, it will still be a green-screen application, and the team must work within those confines. It can be frustrating for some when technology appears to be getting in the way of a great experience. However, working within such confines is part of the skill of a great UX practitioner. Technology has other considerations, as well such as time and budget. Today, almost anything can be done through software, but the time and money required to do it may simply make it unrealistic. Aiming for a great UX is not about ignoring everything else, but balancing considerations and embracing them to craft the best experience possible.

THE UX TEAM SHOULD APPRECIATE TECHNOLOGY AND DEVELOPERS

In the UX Lifecycle, developers do not create the workflow or UI. However, that does not mean technology is not an important consideration. A software project is a balance between the desired outcome and what is technically required to achieve it. It is critical for UX teams to appreciate the challenges, limitations, and obligations of the development team. UX is a priority, but it is not the only consideration of the technical teams.

Examples of such issues are

- The underlying technology may require a significant amount of work to deliver a particular aspect of the desired experience. The business may judge that the cost in time and money is too high, and so that part of the experience needs to be reworked.
- Creating a particular experience may require significant changes to the way in which an existing application is technically structured (the architecture). This could be a significant undertaking in terms of time and money.
- Different applications share certain elements (databases, code, and architecture). While the UX team may be focused on a single application, the technical teams have a responsibility to other parts of the business.

The development team should always look for ways to deliver the best possible experience, but it must be appreciated that there can be issues which hamper this. Sometimes, this will require technical knowledge within the UX team. It was a common experience in the early days of the Web that designers would design websites that were impossible to build with the available technology. This was primarily due to traditional print and typographic techniques not translating well to the Web paradigm. The result was a wasted effort by the designers and then conflict between design and development over why it was impossible to deliver. This does not happen today, as the disciplines should have an appreciation for each other's basic aims and limitations.

The UX team is inevitably creating an experience that will be delivered through software and the technology involved is integral. One cannot create a great iPhone app experience without appreciating the native experience of the device's operating system. The team, especially the UX and UI designers, need to incorporate the technology into their own work.

Developers' Skills Are Critical for Creating the Experience

Developers cannot code everything. Their skills must be accounted for when decisions are made about technology and how that affects UX. Today, most applications are built using Web technologies such as

HTML and JavaScript. According to a Stack Overflow 2022 Developer Survey, these technologies are the two most popular technology skills. [StackOverflow00] Before, many applications were built using traditional programming environments (such as Java, C, or Visual Basic) that some may believe led to inherently poorer experiences.

Many development teams, however, will have been hired for their skills in these traditional technologies and may not have certain skills or experience. If the UX team attempts to dictate the technology, the result in terms of time and cost (hiring or re-training) could be prohibitive. Instead, UX should be (and can be) applied regardless of the technology used.

Developers Need to Be Friends, Not Enemies, of UX

No matter how good the UX team or process, the final experience is entirely dependent on the development team that builds the application. It is not unusual for development teams to have reservations about UX. Typical concerns include

- a feeling that some responsibility is being taken away from them if they were responsible for the application's original UI
- a fear that the UX process is going to slow down and interfere with development
- a concern that implementing UX is going to have a negative effect on the KPIs by which they are judged
- a sense of being criticized if the UX that is now being considered "bad" was originally created by them

It is important to note that having a friendly, cooperative development team is beneficial to delivering great UX. Delivering the best experience often requires a significant amount of technical optimization, and the developers will need to do this. Therefore, taking the time to understand and address any concerns they have is time well spent. The greater challenge is often a cultural tension between the UX team and the development team(s). This tension often stems from a lack of understanding of each other's needs. If developers feel that technology issues are not appreciated or ignored by the UX team, then there will be friction.

Developers: Great Proponents

When developers get to see the benefits of UX, it can create enthusiasm for the process. Developers can appreciate how their work impacts user satisfaction or benefits the business. This is a good reason why the results of UX, like the research, should be available to all.

Education about UX forms a significant part of engaging the development teams. Once developers buy-in, they often become strong UX evangelists. In some scenarios, developers become part of the education process with regards to future development teams. This especially helps to manage cultural conflict.

UI CODE OPTIMIZATION

When building for the Web, optimizing front-end (or UI) code is known as *Front-endOps*. It is an important task in optimizing the performance and integrity of the UI across different browsers and operating system environments.

Many new UI frameworks use a variety of technologies (such as HTML, CSS, SASS, LESS, and JavaScript). Webpages have code that addresses analytics, A/B testing, and device optimization. All of this code needs optimizing prior to deployment to deliver the best possible performance. UX developers can perform that task, freeing the development team for other kinds of work.

HARMONY BETWEEN UX AND DEVELOPMENT PROCESSES AIDS SUCCESS

A UX process should work with any given development process, and UX works particularly well with Agile. This is unsurprising, given that both are essentially iterative improvement processes. Agile, unlike Waterfall, gives UX teams the ability to evolve an application's experience on a regular basis. Agile enables more frequent updates to an application, and that can give users an impression of responsiveness to their needs, which is part of the overall experience.

Agile involves short periods of development called *sprints*, which result in new usable capabilities being delivered frequently. Combining Agile with the UX Lifecycle results in regular software improvements. (Compare this to using Waterfall, where delivery cycles are typically months in length and so any improvements take longer.)

The concern that development teams often have in Agile practices is that UX requires a significant degree of pre-planning and preparation work. This can look like Waterfall with its long requirements-gathering and planning phases. However, this should not be a problem, as the UX process can run one or more sprints in advance of the development team. Agile does involve precise planning, and so such concerns highlight a larger Agile implementation issue.

UX is inherently an iterative process, and Agile is an iterative development approach. In Waterfall, the UX process should still be iterative, but will need to run multiple cycles within the initial, larger, and longer requirements and specifying phase.

Once development is complete, the Build stage comes to an end.

MOBILE UX

Inside organizations, mobile is perhaps the strongest motivator of UX. This is because mobile devices are far less "forgiving" than traditional desktop PCs: The form factor is small with limited space; entering content is awkward and involves using touchscreen keyboards; network connectivity can be slow and intermittent; users are likely traveling, have limited time, and are probably engaged in some other activity. Unless the experience is excellent, then the app or website is almost a waste of time, money, and effort for the organization that built it. This makes mobile UX challenging, but fundamental to an app's success.

There are organizations offering a desktop application with a poor UX that invest in a high quality UX when they create the mobile version. They know that while users may persevere with the desktop version, they will quickly give up on the mobile version. Mobile projects also tend to be small, especially compared to traditional desktop applications or websites, which makes them ideal for trying new processes and skills. They involve fewer employees, have smaller budgets, and the results can be demonstrated more quickly.

This makes mobile projects a useful way to introduce UX into an organization. Whether the project is B2C, B2B, or B2E, its success will largely depend on the experience it provides. Whatever the target audience, many of the business case drivers outlined in this book still apply and, if anything, are even more relevant. Increasing sales, boosting adoption rates, and increased user productivity are all relevant to mobile apps.

While the target endpoint (mobile device) may be different, the principles of UX and the UX Lifecycle are just as applicable. The UX Lifecycle is not designed to be specific to any particular type of experience, and so can be applied to mobile apps or websites as well as to desktops or other types of devices.

KEY TAKEAWAYS

Development teams have a vital role to play in the UX process. The relationship between the development and UX teams is critical. Be mindful of the impact that UX can have on their processes and culture, and seek to win them over (developers make great UX evangelists).

UX is not commonly a skill of software developers, even though they may be responsible for it in some environments. Often, the development process separates developers from users. KPIs are usually focused on delivering functional requirements and not how those requirements are delivered.

UX is a balance between the best experience and other considerations, like the technology used. The UX team must find the balance between these considerations. They must appreciate the challenges, limitations, and other concerns of the technology and development team.

UX should be ambivalent to the software development methodology used by the organization. With development support, UX projects will have better outcomes, ensuring a minimum of disruption to the development teams' processes.

REFERENCES

[StackOverflow00] Stack Overflow. *Stack Overflow 2022 Developer Survey*, available online at *https://survey.stackoverflow.co/2022*, 2022

CHAPTER

9

UX LIFECYCLE STAGE THREE: MEASURE

DEPLOY, VALIDATE, AND REPEAT

In this stage, the application is checked against the original workflow and UI design, any pre-defined metrics, or standards (such as performance or accessibility), and then deployed. Following deployment, the outcome is further validated with users and against the business's original goals to ascertain the ROI. Much of this work could be considered research, and many of the outputs will feed into the next Discover stage. The UX team would typically own this stage of the cycle.

As demonstrated in the Discover stage, a significant amount of research is done to design the best possible workflow and UI. This is to deliver the "ultimate" solution in terms of efficiency, effectiveness, and satisfaction, which leads to the best business outcome. The use of the term "ultimate" here should be considered as being relative to the process of the UX Lifecycle. Inevitably, issues will be found with this version of the workflow and UI, which will then be addressed in a later iteration of the UX Lifecycle.

[There was a] 22% improvement in each iteration where UX was implemented as an iterative process. [NielsenNorman00]

CHECK THAT WHAT WAS INTENDED WAS DELIVERED

Once development has taken place, check that the output from the Build stage matches the input from the Discover stage (allowing for any changes made in the initial phase of the Build stage). Often, and for various reasons, the output is not 100% of what was intended. Any variances or deviations can impact overall success. Whatever the development methodology used, a checkpoint (or "tollgate") should be built in before deployment in which the UX team validates what has been developed. There should be an additional time allowance to make any necessary changes.

There can also be issues that arise out of development that were not or could not be identified earlier. Collaboration between UX and development teams is critical to resolving such issues. Some examples are as follows:

- A particular database responds slowly, and so the application needs to address that issue. If poor performance cannot be changed technically, then the UX team may decide to implement a change in the experience or not make a planned change. The application should not be deployed with serious unresolved UX issues.
- A further validation process is checking pre-defined requirements. One example might be predefined benchmarks like maximum load times for Web pages or data. If it were pre-determined that no user should wait longer than two seconds for the application to respond to any given action, then that should now be tested.
- Accessibility, whereby a website may need to meet accessibility standards such as those outlined by the Web Accessibility Initiative, is important. There are tools for testing this type of criteria.

Include UX Testing in the Overall Testing Plan

There are specific types of tests for UX, most well-known are User Acceptance Testing (UAT) or Usability Testing. These are different types of tests for identifying different types of problem. UAT focuses on whether the application works from a functional perspective, whereas usability testing focuses on user behavior.

As with any functional tests, these return either a "pass" or "fail." Either an application meets a certain accessibility requirement, or it does not. If any tests fail, then further work needs to be done. UX should help to specify these tests, but they are owned by the development team as part of their testing regime.

In the modern application environment, the underlying architecture can be complex. For example, software can have numerous dependencies on other systems, sometimes owned by third parties, which can impact the reliability and performance of the application. Testing the performance of an application can mean testing a sizeable and complex technology stack. There are tools available for this and they are increasingly marketed as UX testing tools. They usually check performance or reliability, which are elements of UX, but are by no means comprehensive. Testing should include both the testing of code and UI, again something which there is an increasing proliferation of tools to do. Many of them are cloud-based and low cost.

UX TESTING METHODS

User Acceptance Testing (UAT)

In these tests, the application is checked to make sure that it does what it is functionally supposed to do from a user's perspective. For example, consider a requirement that is to enter two numbers and then click a button to have them added together, displaying the result. Then the tests would check the following:

(a)   Can the user enter two numbers?

(b)   Does the button add them together?

(c)   Is the result correct and displayed?

Typical testing in software development checks the individual code functions that enable these requirements at the code level, but UAT ensures that they work from the user perspective.

Usability Testing

Whereas UAT is interested in whether the application meets the functional requirements, usability testing looks at user behavior. Taking our previous example, the application may generate the correct result, but usability tests might consider the following:

(a)   Does a user know to enter two numbers?

(b)   Do they know where to enter the numbers?

(c)   Do they know to click the button and what is expected to happen?

(d)   Is the result obvious?

HOW DEVOPS HELPS TO DELIVER AND VALIDATE UX

Once the application has been tested and validated by the development and UX teams through testing, it is ready to deploy. Deployment often means moving from a technical testing environment to a live production environment. These can have significant differences. For example, the volume of data in a test database can be substantially less than in production. That can impact the performance of the application when reading or writing data. Therefore, any benchmark testing should be repeated against the live environment. In many development teams, this type of testing in production already occurs, and so the UX tests just need to be included.

In large organizations, the relationship between the development teams and the teams responsible for providing and managing the live production environment (often known as *operations teams*) can make deployment challenging. Technical differences between the development and production environments can mean code that works in development does not work (or work the same) in production.

DevOps can help address these challenges. DevOps is usually found in agile environments where frequent updates are being made to the application that benefits from having an efficient process for delivering those changes to the user. The DevOps process is about moving an application from development to production as quickly and smoothly as possible and then measuring it for any potential problems. Problems are traditionally technical issues, but can include checking for usability issues. The latter is seen by some as a feedback loop through which analytics data is reported back to the development teams and business. This helps to identify problems that can then be addressed in a further iteration.

Be Cautious of DevOps Pitches That Include UX

Some vendors of DevOps tools have bundled UX into the definition of DevOps. The feedback loop is seen as being the user experience process. As we have outlined, UX is far more than this small aspect. In fact, deployment is just part of a single stage in the overall UX Lifecycle. Simply providing data is of little use without the skills to interpret that data and the processes to create a useful tool as a result. Therefore, one should be wary of such views pertaining to DevOps and UX. DevOps is about smoothing the path from software development to software deployment, which is an important step in the overall application lifecycle.

DevOps does speed up deployment, and this can contribute to a better user experience. Users can often feel that an application is more responsive to their needs if updates frequently address key pain points or simply improve the experience. Should problems occur in the software, then being able to rapidly correct them helps to regain users' trust and minimize reputational damage. This is worth noting because it highlights the breadth of what forms the experience of using an application.

WHY ARE UX LIFECYCLE STAGES MORE IMPORTANT THAN SPECIFIC ACTIVITIES?

The UX Lifecycle is designed in three stages that can be repeated for additional value. Within those three stages, there can be many different types of UX activity, from varying forms of research to differing levels of analytics reporting. The UX Lifecycle does not prescribe any particular activities for several reasons:

- Budget – Organizations and projects may have different scales of budget and so teams can choose the UX activities that best fit the money available.
- Application – The appropriate UX activities can be selected to address the software being created, whether that is for a website, mobile app, wearable app, or something else.
- Human resources – Some organizations may have a large UX team while others may have just an individual. Given the scale of the human UX resources, the optimal UX activities can be used.

USERS ARE CRITICAL, AND SUCCESS IS THE FINAL DELIVERABLE

The Measure stage does not just measure the application in a quantitative sense (such as response time or number of crashes), but also further qualitative measurements. This means soliciting views from users as to the level of improvement in the experience. Finally, the business objective(s) that was the purpose of this cycle of improvement must be measured.

The original objective for the business may have been to reduce the number of abandoned shopping carts, decrease the time it takes employees to complete a given task, or improve the quality of data being collected through a certain process. Now this outcome should be measured and compared to previous measurements to ascertain how successful the UX process has been. UX should result in measurable benefit to the business.

Measuring can be done in various ways. Having a capability built into the product can track metrics such as task times, number of people completing a task, or number of errors that occur. There are measuring tools for finding this type of data, and these tools can be integrated into applications, regardless of where they run.

There is also insight that can be derived from data that the application inherently captures. If the number of abandoned shopping carts must be measured, then there is probably a database that stores data for every shopping cart created. That can be queried to see how many customers completed the purchase.

Finally, some of the methods from the Discover stage, such as the monitoring of users, user groups, and surveys, can be utilized. Many of these quantitative metrics can be fed into UX dashboards that show, at-a-glance, pre-defined KPIs so that stakeholders can see how effective the latest iteration of the UX Lifecycle was. Once success has been measured, the cycle can begin again.

KEY TAKEAWAYS

The experience that was created in the Discover stage may have changed during the Build stage, and the output needs to be validated. Development can result in inadvertent or necessary changes to the original experience, and they need to be addressed before deployment.

Development needs to include new UX-driven tests, such as usability testing or user acceptance testing, in their overall testing plan. The different environments in which an application operates (development, test, and production) can affect the experience differently, and so tests should be repeated in all environments.

Once the application has been deployed, further research should be done by the UX team to establish how well it has addressed the original objective(s). Some of the data gathered should enable the ROI to be calculated. Data from this stage will be used in the next Discover stage.

Too often, organizations invest in UX, but not in the ability to measure the outcome. This will usually require new data capture capabilities configured according to the objectives set for each iteration of the UX Lifecycle. Measurement is critical to the success of the software.

REFERENCES

[NielsenNorman00] Nngroup.com. *Parallel & Iterative Design + Competitive Testing = High Usability | Nielsen Norman Group Report*, available online at *https://www.nngroup.com/articles/parallel-and-iterative-design/*, 2011

10

MATURITY MODEL

MATURITY LEADS TO MORE EFFICIENCY AND LESS FRICTION

The ideal scenario for implementing UX is to have all the necessary skills and processes in place from the beginning. This can be difficult for most organizations, and some enterprises may want to try UX on a small scale. Until the ROI is proven, and the rest of the stakeholders are convinced about the benefits and educated about the process, UX may need to operate in a limited way. Fortunately, the UX Lifecycle is not dependent on size or scale.

Any organization will need to balance risk and reward in the early stages, if only to get the necessary budget approved. Once the benefits and model are established, an organization may then choose to grow its UX capability.

This can be done in a variety of ways, such as from an organic approach, in which new resources are brought in when needed, or in a more deliberate way through a sudden growth in capability. For example, GE built a UX center of excellence, and Accenture acquired a sizeable company to operate at a significant scale. Regardless of the approach used, the people and processes involved will need to scale up.

The UX Lifecycle can be applied to any scale of UX implementation. However, one of its unique benefits is that it works on a larger scale. This is often a challenge for UX teams who work well on small projects one at a time, but find it difficult when dealing with multiple projects across many business functions and with numerous teams. In addition to scale, there are issues that emerge over time.

> [Estimates indicate] $1.3 trillion per year could be saved by business globally through improving the usability of intranets. [NielsenNorman00]

Best Practices and UX Debt

Working practices and other reusable assets will accrue over time. For example, a UX guide may be developed that contains the best practices, such as UI elements or target performance metrics and standards. Such a guide enables less skilled people to apply some degree of UX to projects (perhaps smaller ones where the budget does not cover a UX team). It also helps to keep multiple UX teams in sync. IBM created its own design language that enables multiple teams to work on different applications in different geographies while maintaining a consistent experience.

Another issue is UX debt, which is modeled on the concept of *technical debt* that development teams typically have. *UX debt* is where elements of an application's experience do not fully comply with the approved UX. This may occur through the omission of parts of the application in the UX process or parts of the project do not meet the deadlines as the UX process evolves. This can happen for numerous reasons, but it needs to be addressed. An efficient way of managing UX debt is to have a UX resource, perhaps a UX developer, dedicated to examining the applications, identifying the debt, and then developing a program of work to correct it. By eliminating bad UI (debt), you can only ever create an average experience. It is only when paired with improving the user workflows that you can create a great experience.

Signs of a Maturing UX Capability

A sign of maturity is the ability to deliver shorter implementations of the UX Lifecycle alongside longer ones. An example of this can be found in A/B testing. This is where research identifies potential changes that could be divided into two possible solutions (A and B). These potential solutions are then applied in a production environment quickly, without going through a full development process (or even the development team). The results can then be measured quickly, and the best solution is used in the next full development iteration. There are tools that enable this type of capability, but the teams need to be empowered to use them.

A further step is to provide UX reporting to the various stakeholders. The development of dashboards showing performance and usage metrics along with other types of research and measurement data are growing in popularity.

WHY UX FAILS

Organizations are not always completely unaware of UX nor do they always do nothing to address it. In many cases, it is a consideration in the application development process. That does not mean it is always successful, and problems within UX processes can result in less-than-successful outcomes. This is a reason why measuring the outcome of UX is so important. One should be alert to issues that can lead to a failure of UX.

There are a few reasons why UX might not succeed:

- failure to understand UX
- failure to educate the project stakeholders and business
- failure to research and engage with the users
- failure to define the problems
- failure to spot UX problems
- failure to validate UX work before releasing the final product
- failure to pay off UX debt and so let parts of the application experience deteriorate
- failure to measure the ROI

INDUSTRIALIZED UX FOR OPERATING AT SCALE

UX has typically been done in small teams focused on a single project. This is a common approach in creative agencies, where a team of a few designers and developers will create a solution that may have only a short lifespan (such as a seasonal campaign). This means that many of the traditional UX practices and processes have evolved out of this type of environment. UX at scale is something that large vendors like IBM, Hewlett Packard Enterprise, and Google are addressing. When Larry Page became CEO of Google, one of his first acts was to outline the intent to create a unified and consistent experience across Google's many different products.

Large businesses often function in product silos that do not reflect the world that their users live in. A single user may use multiple products, and they want an experience that spans them all. Adobe built out its suite of creativity products through acquisition and then invested in giving them a unified experience, just as Microsoft has with its Office suite. In that way, a user can move between products far more easily.

Scale Means Greater Collaboration and Formalization

In situations where the number of people or applications involved or the typical application lifespan are longer, this traditional approach is challenged. The result is that there needs to be a greater reliance on formalized process (such as the UX Lifecycle) and the tools to manage it. Breaking down skillsets into specific roles is one way to help UX teams work across multiple projects.

More importantly, there needs to be a form of oversight or governance to the UX function. This can take the form of a UX strategist or UX director who has responsibility across all UX activities. In a very large-scale scenario, there can be multiple UX teams and yet the key tenets of UX require all of them to be operating in sync at the process level down to specifics of implementations (examples are having shared UI conventions or performance metrics across all applications).

Scale can also mean more stakeholders, such as management, product owners, users, or others. When an application's lifespan is long, such as that of an Enterprise Resource Planning (ERP) system that can exist for many years, there needs to be consistency over time. Steering groups made up of key stakeholders can effectively manage consistency. UX involves collaboration and process, so when operating UX at scale, all members of the UX team must communicate and share research, best practices, and guidance.

Once formalized processes have been created with the necessary education, skills, and tools in place to support them, an organization might consider creating a center of excellence or competency center. These are central hubs responsible for creating and disseminating best practices and supporting implementation across the organization.

The Future Holds Diverse Experiences

The types of application are evolving rapidly. The desktop application used to be the staple of an organization, but now applications can be delivered via Web browsers, mobile devices, wearables, and Internet of Things devices (such as cars, thermostats, and refrigerators). UX will be critical in the case of virtual reality or augmented reality applications. An organization will need to make strategic and tactical decisions about how to handle this from a UX perspective. Having a UX capability already in place will help.

Consider the importance of UX principles, such as consistency and familiarity, and how these need to be applied across a diverse application landscape. A user should not expect to get a different experience from using different mobile apps because they were developed by different UX teams. The same applies for when users move between using desktop and mobile apps. Microsoft recently introduced the Universal App in Windows 10, which enables the same application to run across multiple device types. Creating such an experience comes with new challenges. The speed of change in technology is such that it is better to act now with respect to UX rather than wait until the competition has surpassed you.

WHAT SHOULD I EXPECT AT THE START THE UX JOURNEY?

There are challenges to overcome after you have decided that UX will deliver value to your organization, made the business case, and brought in the appropriate resources. The first is that UX cannot immediately deliver tangible results. We have outlined how critical research is to UX, and a reasonable amount of time will need to be allocated for the initial Discover stage.

In the beginning, the UX team will potentially have limited knowledge of the application, business domain, or organization. They will need time to investigate all of these and will need to understand

- the purpose of the application and who in the business has ownership of it
- who the main stakeholders in the application are
- who the users are
- how UX is currently created in the organization

Once these areas have been addressed, the team will then need time to understand the scope of their work. Applications differ in size based on capabilities and complexity, and there may be a grand vision that needs to be achieved in smaller, incremental steps.

There will also be the challenge of working out how the team will work with other stakeholders, most notably the development team, and then to put in the processes needed to facilitate UX.

Addressing all of this is critical to delivering a great experience, and so it is important to invest the time in doing it. Even after the team begins to deliver results, it takes time for these changes to reveal their true impact. Good UX should have some positive results immediately, but change can be challenging for all stakeholders, not just users. Therefore, it can take more than one iteration of the UX Lifecycle to truly begin realizing value that everyone can appreciate.

CREATING A MORE EFFICIENT IMPLEMENTATION WITH A UX PLATFORM

The UX Lifecycle discussed in this book is a series of stages with various deliverables throughout. Consider how research data is produced by the research team and delivered to the UX design or architecture teams. They provide wireframes to the UI designers, who in turn provide designs to development teams, often as flat image files. Developers then turn these designs into code. This process works well for a single project or one that is short-term. However, it can create challenges or inefficiencies for larger or longer projects.

An example is the handover between UI design and development (between the Discover and Build stages). The development team receives an image or images of the screens that they need to build for the user workflow. Developers must use code to create screens based on the designs.

When the design changes in the next iteration of the UX Lifecycle, the development team must identify the changes within the new image files provided by the UI designer, and then make the required code changes. If anything within the design changes during the Build phase, then the UI designer must make the necessary changes in their image files to remain in sync for the next iteration. Over time, this can require considerable work from developers and creates potential for numerous types of errors. An organization should attempt to create a situation where different disciplines and processes can operate in a more efficient and frictionless way.

[There was a] 161% increase in click-throughs for software company Veeam by changing a single word. [Econsultancy00]

Enable UX Lifecycle Stages to Flow Smoothly

To improve the flow of work in the UX Lifecycle, the team should move away from asset-based deliverables (such as documents, images, and files) toward a more integrated means of design and development. The team could develop a working set of coded common UI components. These components can be managed and maintained by the UX team, perhaps using a UX developer.

UI designers within the UX team can then "design in code." Instead of designing in graphics tools (such as Photoshop), they can use the working components to put together a new UI. These same components can be used to create prototypes more easily and quickly for user testing. The development team then uses the same components when building the application. They do not need to spend as much time translating images into code, and the integrity of the components design and experience will be maintained.

If an aspect of the components needs to change, then this can be done by the UX team and automatically reflected in the design, prototypes, and development environments. Such an approach reduces friction as a project moves through the stages of the UX Lifecycle. It also removes unnecessary effort on the part of design and development teams, and reduces the potential for UX debt, especially where changes are made to UI conventions either because of new research, requirements, or changes to the brand.

These components can be shared across multiple teams and applications, and so operating UX at scale becomes easier to manage and ensures the main principles, such as consistency, familiarity, and expectation, are maintained across all applications.

This mature approach can be called a "UX platform" on which much of the UX Lifecycle can be built. When operating across multiple applications and over time, this can have considerable benefit. With all teams working on the same platform, the main principles of UX (such as consistency and familiarity) are easier to apply and manage. It brings a greater level of efficiency and effectiveness to the teams involved.

BEYOND METHODOLOGY: A UX PHILOSOPHY

Perhaps the ultimate sign of maturity is when UX becomes more than process and is instead a philosophy. This is the point at which all the stakeholders inside and outside of the organization share a common ambition to deliver a great experience to users. This is not about making everyone UX experts, but instilling an understanding of the benefits and how they are released (such as the UX principles).

In this scenario, there would be no need to remind someone fighting for their opinion in a design meeting to check the research, because everyone would inherently be driven by the research. Developers would automatically appreciate every nuance of the experience and how their code affects the user and the business. Once UX becomes a philosophy of an organization, it becomes integrated into everything they do.

However, such a business culture would also have a deep appreciation that perfection is never attained, and everyone can always do better to deliver a better experience for the user.

KEY TAKEAWAYS

UX has traditionally been done within smaller teams, such as those inside design agencies. When operating at a greater scale, more formalized processes and tools are required. The UX Lifecycle does not dictate or depend upon a certain scale, but can support large scale environments with multiple complex applications and multiple UX teams.

Signs of UX maturity within an organization include the ability to run shorter UX Lifecycle iterations within larger ones; moving towards different disciplines, and processes operating more efficiently and with less friction; dashboards showing real-time metrics related to the UX; and recognizing and addressing UX debt.

A UX platform should move away from asset-based deliverables being passed between stages towards more integrated research, design, and development. The use of shared component libraries and the ability for a UX team to "design in code" are ways to achieve this.

Ultimate maturity occurs when UX becomes a philosophy within the organization. Stakeholders embrace UX thinking at all levels and an appreciation for delivering the best user experience becomes part of how the organization works.

REFERENCES

[NielsenNorman00] Nngroup.com. *Intranet Usability: The Trillion-Dollar Question*, available online at *https://www.nngroup.com/articles/intranet-usability-the-trillion-dollar-question/*, 2002

[Econsultancy00] Econsultancy. *Conversion rate optimisation: eight case studies that show the benefit of UX testing*, available online at *https://econsultancy.com/blog/63984-conversion-rate-optimisation-eight-case-studies-that-show-the-benefit-of-ux-testing/*, 2013

11

CASE STUDY 1: BIG POWER COMPANY

Big Power Company is a large organization that supplies power to homes and businesses. Many of its thousands of employees are tasked with maintaining and repairing the vast infrastructures that its business requires.

They decided to provide their field force with tablet devices running several of the company's proprietary applications to make them more productive and reduce job costs. The savings would come from a reduction in paper and print costs; less time spent by office workers preparing the job information (especially with integration with existing IT systems); and less time spent processing paper-based forms returned by the work crews at the end of each day.

The company's IT team built the applications based on requirements provided by the relevant business units. Each worker was provided with training in how to use the device and applications. The main application is one that manages jobs by providing job information, including materials such as maps and plans.

Identifying the UX Problem

After launching the applications, the business saw many of the anticipated reductions in paper/print costs and the time spent by office staff preparing job information. However, there was no reduction in job times and in some cases, there was an increase. Members of the field force were frequently calling IT support and other colleagues to get information that was available via the applications. Instead of getting real-time updates on progress, the business was still being supplied completed job information in unorganized paper form by the work crews at the end of each day. Having made an investment in hardware and software, the business was not obtaining the desired return.

To try and understand the problem, some anecdotal evidence was gathered that suggested work crews were finding it difficult to use the applications for accessing and entering information. Based on feedback from the business units, IT made changes to the applications, but the problems were not resolved. The applications provided all the requested functionality and worked correctly. One of the senior IT managers decided that the problem was the user experience of the applications.

Making the Business Case

As the company had no skills in user experience, the senior manager knew that an investment was going to be required to acquire the necessary capabilities. To justify this expenditure, she proposed the potential savings could be realized by improving the primary job management application. Starting with just one application seemed to be a sensible way to introduce the concept of UX into the business and test what value it could deliver. The company leadership examined three areas: a reduction in job times; fewer support calls to IT; and less time spent by office staff processing paper-based forms.

By using calculations based on employee costs among the three roles involved, they were able to project the savings. For example, they identified the following budget items:

- An IT support employee costs the business $26,000 per year (48 working weeks).
- Support call logs showed that they dealt with an average of ten calls a day from work crews.
- The average call length was ten minutes.
- The goal was to reduce the number of calls by 50%.

For each IT support staff member, that would mean the savings would be as follows: ($26,000 / minutes in a working year) x (10 calls x 10 minutes) = 22.6 cents per day on work crew calls

22.6 x 5 days x 48 weeks = $5,424 per year on work crew calls

3,744 / 2 (50% saving) = $2,712 saved per year per support staff member

With a total of sixty in the support team this would mean an annual saving of $162,720.

Looking across the savings to be made in all three roles, there was enough money in the budget to justify hiring a UX expert. Armed with this business case, the senior manager sought buy-in from the CIO, who agreed with her proposal. Speaking with several potential candidates it became clear that a separate UX design role would also be needed. The business case was adjusted and approved.

Getting Started

A UX expert with several years' experience in delivering UX programs was hired. His skillset encompassed several areas of UX, including research, design, and team leadership experience.

The first task of the UX manager was to bring together the stakeholders across the business units and IT. He presented to them what UX was, the type of benefit it could provide to their functions, and how these benefits would be realized through new processes. He also addressed members of the development team to provide an understanding of what UX would mean for them and to offer reassurance regarding their role.

He then held several workshops with members of the work crews. This was to understand how they used the job management application and the challenges that they were experiencing. Based on this research, the UX manager was able to report back to the stakeholders on why the users were struggling with the application and provide general guidance on what could be done about it. This reinforced the projected savings that had been outlined in the business case.

The UX manager then hired a UX designer, and the project began. At this stage, the UX team was operating within the available budget and focused on a single application. Based on research, the application was redesigned. This included recreating some of the core workflows, as the problems were not just with the UI, but the flow of tasks within the application. A prototype was tested with the users in a workshop environment and further refinements were made.

Development

Once the new application experience was agreed to by the primary stakeholders, it then moved into production. The UX designer and manager worked closely with the development teams to ensure that the intended UX was delivered. The development process took longer than anticipated, as the development teams struggled with some aspects of implementing the new UX. The new application was then delivered to the work crews.

Over the first month of usage, there was a noticeable drop in average job times, calls to support staff, and the amount of paperwork being produced. The second month saw a greater fall as the teams became increasingly familiar with the new application. The savings came close to those outlined in the original business case. Further workshops were run with the users, and another round of changes were made to the application. This time, the savings were on target with the business case.

Scaling the UX Capability

Inspired by this success, the organization decided to extend the UX work to more mobile device applications. To help scale UX delivery, the existing team created a UX platform based on a core set of components that could be reused across applications. A UX developer was also hired to work within the various development teams to avoid any further delays in this part of the process. They also built out the platform components in a way that could be used in the design, prototyping, and development stages. This made the process of design and development far more efficient.

As more applications were included in the UX process, the benefits were repeated. Over time, the number of support calls was reduced to almost zero, training time for the work crews was cut by 50%, and all use of paper in the job processes was eliminated.

Today, all business teams involved in the creation, management, and support of applications are well versed on the principles of UX, familiar with the benefits, and understand the research and research processes. This means that UX is considered one of the priorities in any discussion related to the applications.

12

CASE STUDY 2: MIDSIZE SOFTWARE COMPANY

Midsize Software Company is an Independent Software Vendor (ISV) with an application for managing stock inventories. The company has been around for many years and recently migrated its traditional desktop/server application to the cloud. It is now available to customers through a Web browser. This allowed the company to move from traditional licensing to Software as a Service (SaaS) with a monthly subscription. Its customers are mainly small and medium-sized businesses. Midsize Software is a small company of mainly developers and a few salespeople, but the new SaaS model allows it to sell internationally.

Identifying the UX Problem

Over a period of months, sales of the product began to decline. Even when the sales team was engaging directly with larger buyers, there were fewer customers than before. They believed that some existing customers and potential new ones were choosing competing products. This was despite the product having the most complete set of features and maturity in the marketplace.

Members of the sales team began asking some of those customers who chose an alternate product about the reasons behind their decision. Almost all of them indicated that they felt competing products would be easier for them (and their employees) to use and that they "looked more modern." The business leaders decided that they needed to improve the interface of their application to address these concerns.

Making the Business Case

Being a small technically-led business, the leadership realized that new human resources were going to be needed to design a better UI. To justify the cost, they had to consider the projected loss in revenue based on the slump in sales that they had witnessed. The current sales slump was 20% compared to previous comparable periods. This translated to a loss of monthly income of $4,500 and a yearly loss of $54,000.

The decision was taken by senior management to contract a UX consultant. As part of the selection process, several candidates provided management with insight into UX and the skills, processes, and potential benefits beyond just their initial sales concern. For example, they learned that UX could reduce support requests and potentially assist with future product direction by understanding better what users value and need.

Getting Started

The successful provider had skills in UX research, a familiarity with design and development, and the processes involved in implementing and embedding UX into a development environment. The first task was to interview users, which was challenging as they were customers. Instead, the analytics built into the application were used to identify different user journeys and where users were either not following typical workflows or encountering barriers. To gather the best data, some changes were required to the configuration of the analytics tool.

Once a suitable data set had been gathered and analyzed, the consultant was able to provide the company with a report outlining the issues with the existing product. Several solutions were also suggested. Management decided to continue with the project, and the consultant worked with their own UX designer to create a new experience.

Development

The consultant then worked with the development team to implement this in the product. There were concerns among the developers that the UX process was going to negatively impact their own processes and ability to deliver.

Once complete, the new version was tested with some of the same group of users that took part in the original interviews. Based on their feedback, further changes were made, and the product was released to all customers.

The results were viewed positively among many customers, although there were some initial concerns due to the change. The sales team immediately saw a resurgence in sales, although not as much as the original slump. This was enough to convince the company that the exercise had been worthwhile. After seeing the positive results and feedback from customers, the development teams also became convinced of the value in the UX process.

Scaling the UX Capability

The decision was made to hire a UX expert with research and design skills. The company leadership realized the importance of research, but also needed in-house design capabilities and did not have the budget for two roles. The recruit worked with and improved upon the processes installed by the consultant. They also integrated their processes tightly with the development teams and invested in educating the developers about UX.

With only one dedicated UX practitioner, it was critical to have development teams with a strong understanding of UX. A UI guide was created to enshrine certain UI principles in a way that helped to ensure consistency, minimize UX debt, reduce design times, and assist the development teams.

Now the company can continue to improve on the application experience while also adding new capabilities that are motivated by users, rather than suppositions on the part of the management. UX has begun to become part of Midsize Software's overall culture. Over time, sales not only returned to previous levels, but increased as customers viewed the company's application as better than the competition with respect to capabilities and experience.

*I*NDEX

www.ingramcontent.com/pod-product-compliance
Lightning Source LLC
Chambersburg PA
CBHW071221050326
40689CB00011B/2401